A. J. Kfoury

An Introduction to Formal Language Theory

With Contributions by James Pustejovsky

With 61 Illustrations

Spring-Verlag New York Berlin Heidelberg
London Paris Tokyo

Robert N. Moll
University of Massachusetts
   at Amherst
Amherst, MA 01003
USA

Michael A. Arbib
University of Southern California
Los Angeles, CA 90089-0782
USA

A. J. Kfoury
Boston University
Boston, MA 02215
USA

James Pustejovsky (*Contributor*)
Brandeis University
Waltham, MA 02254
USA

Series Editor

David Gries
Department of Computer Science
Cornell University
Upson Hall
Ithaca, NY 14853
USA

Library of Congress Cataloging-in-Publication Data
Moll, Robert N.
   An introduction to formal language theory.
   (Texts and monographs in computer science)
   Bibliography: p.
   1. Programming languages (Electronic computers)
2. Formal languages.   I. Arbib, Michael A.
II. Kfoury, A. J.   III. Title.   IV. Series.
QA76.7.M65   1988   005.13   88-2025

Typeset by Asco Trade Typesetting Ltd., Hong Kong.
Printed and bound by R. R. Donnelley & Sons, Harrisonburg, Virginia.
Printed in the United States of America.

9 8 7 6 5 4 3 2 1

ISBN 0-387-96698-6 Springer-Verlag New York Berlin Heidelberg
ISBN 3-540-96698-6 Springer-Verlag Berlin Heidelberg New York

In Memory of
James Peter Thorne
Linguist and Lover of Language

Preface

The study of formal languages and of related families of automata has long
been at the core of theoretical computer science. Until recently, the main
reasons for this centrality were connected with the specification and analysis of programming languages, which led naturally to the following questions. How might a grammar be written for such a language? How could we
check whether a text were or were not a well-formed program generated by
that grammar? How could we parse a program to provide the structural
analysis needed by a compiler? How could we check for ambiguity to ensure that a program has a unique analysis to be passed to the computer?
This focus on programming languages has now been broadened by the increasing concern of computer scientists with designing interfaces which
allow humans to communicate with computers in a natural language, at
least concerning problems in some well-delimited domain of discourse.
The necessary work in computational linguistics draws on studies both
within linguistics (the analysis of human languages) and within artificial
intelligence. The present volume is the first textbook to combine the topics
of formal language theory traditionally taught in the context of programming languages with an introduction to issues in computational linguistics.
It is one of a series, The AKM Series in Theoretical Computer Science,
designed to make key mathematical developments in computer science
readily accessible to undergraduate and beginning graduate students. The
book is essentially self-contained—what little background is required may
be found in the AKM volume *A Basis for Theoretical Computer Science*.

After an overview of the entire territory in Chapter 1, we present the
"standard material" for a computer science course in formal language theory in Chapters 2 through 5—the various families of languages: regular,

context-free, context-sensitive, and arbitrary phrase structure languages; and the corresponding families of automata: finite-state, push-down, linear bounded, and Turing machines. Along the way we introduce a number of related topics, including closure properties, normal forms, nondeterminism, basic parsing algorithms, and the theory of computability and undecidability (developed at greater length in the AKM volume *A Programming Approach to Computability*).

Much of the American work on theoretical computer science has emphasized combinatorial techniques, but there is a growing appreciation of the role of algebraic techniques in analyzing the syntax and semantics of programming languages (see the AKM volume *Algebraic Approaches to Program Semantics*). We introduce the reader to this area by devoting Chapter 6 to the fixed point approach to the analysis of context-free languages. We then devote Chapter 7 to advanced topics in parsing, which provide the bridge from formal language theory to computer language design, namely the study of LL and LR grammars. We believe that readers will find this an unusually accessible, yet precise, introduction to the subject.

Finally, Chapters 8 and 9 present our introduction to the formal theory of natural languages. We show how an analysis of properties of English sentences (e.g., the relation of active to passive sentences or the need to mark agreement between subject and verb) led to the development of Chomsky's theory of transformational grammar. We then develop alternative schemes for representing natural language, e.g., augmented transition networks (ATNs) and generalized phrase structure grammar (GPSG), and offer some formal analysis of their properties.

The first draft of the last two chapters was prepared by James Pustejovsky, who then collaborated with the authors in producing the version that we now place before you. We are most grateful to him for sharing with us, and thus with the reader, his expertise in both transformational grammar and computational linguistics. With his contributions, we reiterate that the study of languages by computer scientists can no longer be restricted to programming languages, but must include the study of natural languages, as artificial intelligence and the design of natural language interfaces become ever more important topics within computer science.

The book grew out of our teaching of classes at the University of Massachusetts at Amherst and at Boston University over several years. We thank our students for their helpful feedback as we developed this material, and we thank Darlene Freedman, Rae Ann Weymouth, and Susan Parker for their help in typing the manuscript.

October 1987

R. N. MOLL
M. A. ARBIB
A. J. KFOURY

Contents

CHAPTER 5
Turing Machines and Language Theory

CHAPTER 6
Fixed Point Principles in Language Theory

CHAPTER 7
Parsing, Part II

CHAPTER 8
The Formal Description of Natural Languages

CHAPTER 9
Recent Approaches to Linguistic Theory

Introduction

1.1 The First Language

The modern theory of formal languages stems from two sources: the American linguist Noam Chomsky's attempt in the 1950's to give a precise characterization of the structure of natural languages, such as English and French, according to formal mathematical rules, and the development of a formal specification for the computer language ALGOL 60. Chomsky's work sought to describe the syntax of natural language according to simple replacement rules and transformations. Chomsky considered a number of possible rule restrictions, the most promising of which gave rise to a class of grammars known as *context-free grammars*, which generated the "deep structures" upon which transformations operate. Context-free grammars and the class of context-free languages they generate will be the principal object of study in this volume. This is primarily because the development of ALGOL 60, which soon followed Chomsky's attempted characterization of natural language, demonstrated that context-free grammars are reasonably adequate systems for describing the basic syntax of many programming languages. We will also study some issues in natural language processing, given the increasing importance of computational linguistics in the construction of natural language interfaces.

As a first example of a context-free language, we examine the "language" of matched parentheses. This language plays a fundamental role in computer science as a notation for marking "scope" in mathematical expressions and programming languages. In the latter case, **begin** and **end** are often used instead of "(" and ")".

1 Example. Matched parentheses include all legally balanced strings of left and right parentheses—for example, (()), ()(), and (()(())). The following specification describes the language of matched parentheses inductively:[†]

(1) The string () is well-formed.
(2) If the string of symbols A is well-formed, then so is the string (A).
(3) If the strings A and B are well-formed, then so is the string AB.

We can now follow this inductive definition to give a rewriting system—a grammar—that generates exactly the set of legal strings of matched parentheses.

$$(1) \quad S \to ()$$

$$(2) \quad S \to (S)$$

$$(3) \quad S \to SS$$

These three rewriting rules are called *productions*. They tell us "given a string, you may form a new string by replacing an S by either () or (S) or SS." Notice that (3) uses different notation from part (3) of the inductive definition given first. When we identify A and B in the inductive definition as arbitrary matched parentheses expressions, we use different names—A and B—to indicate that they need not stand for the same string. The rule $S \to SS$ really says the same thing, because the two right-hand side S's may be replaced independently. To generate, say, (())(()()), we go through the following replacements:

$$S \Rightarrow SS \Rightarrow (S)S \Rightarrow (())S \Rightarrow (())(S) \Rightarrow (())(SS) \Rightarrow (())(()S) \Rightarrow (())(()())$$

Here, the production $S \to SS$ is applied first, and then the two new S's are independently rewritten to the strings (()) and (()()).

We can summarize the preceding derivation with the notation

$$S \overset{*}{\Rightarrow} (())(()())$$

which we describe as "S derives (in several steps) (())(()())." When we wish to describe $v \Rightarrow w$ in words, where v and w are arbitrary strings, we shall say that "v *directly derives w*." Thus SS directly derives $(S)S$.

As written, the grammar presents two different kinds of symbols: characters that may appear in derivations, but may not appear in final strings; and characters that make up the alphabet of the strings we are trying to produce. The first type are *nonterminal symbols* or *variables*. In this example, S is the only variable. The second type are the *terminal symbols*. Terminal symbols make up the strings of the language that the rewriting rules are intended to generate. Here "(" and ")" are the terminal symbols.

The programming language style for writing grammars, Backus–Naur form, or BNF for short, uses a slightly different notation. (Backus–Naur form

---

[†] The way in which inductive proofs and definitions can be generalized from the natural numbers to more general structures is discussed at length in Chapter 2 of *A Basis for Theoretical Computer Science*.

is named for two computer scientists, John Backus and Peter Naur.) In BNF, variables are enclosed in angle brackets, and $\rightarrow$ is replaced by the symbol ":$=$". Thus, the second production of Example 1 is written in BNF notation as

$$\langle S \rangle ::= (\langle S \rangle)$$

Matched parentheses, then, is our first example of a formal grammar and an associated formal language. We give precise definitions for these concepts in the next section.

EXERCISES FOR SECTION 1.1

1. What language does the grammar

$$S \rightarrow ()$$
$$S \rightarrow)($$
$$S \rightarrow SS$$
$$S \rightarrow (S)$$
$$S \rightarrow)S($$

generate?

2. What language does the grammar

$$S \rightarrow 0$$
$$S \rightarrow 1$$
$$S \rightarrow S0$$

generate?

3. Does this grammar:

$$S \rightarrow ()$$
$$S \rightarrow (S)$$
$$S \rightarrow SSS$$

generate exactly the language of matched parentheses?

1.2 Grammars and Languages

Let X be a nonempty finite set of symbols, or *alphabet*. Let X^* denote the set of all finite length strings over X. This infinite set of strings includes the empty string, which is denoted by the symbol λ (the Greek letter lambda). If $X = \{(,)\}$, then X^* is the set of all finite strings of parentheses, whether they match or not.

A *language* L over X is any subset of X^*. Thus the language of matched parentheses, M, is a language because $M \subset \{(,)\}^*$. Each string w in L has a finite length, $|w|$, with $|\lambda| = 0$. If $w = x_1 x_2 \ldots x_n$, each $x_j \in X$, we have $|w| = n$. We shall sometimes write X^+ when we mean the language of nonempty strings over X. If $w = x_1 x_2 \ldots x_n$ and $w' = x'_1 x'_2 \ldots x'_m$ are any two strings, we may *concatenate* them to obtain

$$ww' = x_1 x_2 \ldots x_n x'_1 x'_2 \ldots x'_m$$

We set $w\lambda = \lambda w = w$.

Now we come to the notion of a grammar. In the matched parentheses example, we considered rewrite rules that involved characters from terminal and nonterminal alphabets. These three elements plus a fourth, a distinguished variable that is used as a start symbol, make up a *phrase structure grammar*, or simply *grammar* for short.

1 Definition. A *phrase structure grammar* G is a quadruple (X, V, S, P) where X and V are disjoint finite alphabets, and

(1) X is the *terminal alphabet* for the grammar;
(2) V is the *nonterminal* or *variable alphabet* for the grammar;
(3) S is the *start symbol* for the grammar; and
(4) P is the grammar's set of *productions*. P is a set of pairs (v, w) with v a string on $(X \cup V)$ containing at least one nonterminal symbol, while w is an arbitrary element of $(X \cup V)^*$. An element (v, w) of P is usually written $v \to w$.

The grammar for the language of matched parentheses may be written as

$$G = (\{(,)\}, \{S\}, S, \{S \to (), S \to (S), S \to SS\})$$

In general we shall use uppercase Roman letters for variables, and either decimal digits or lowercase Roman letters for terminal symbols and for strings of terminal symbols.

2 Definition. Let G be a grammar, and let y and z be finite strings from $X \cup V$. We write $y \Rightarrow z$ and say y *directly derives* z, or z is *directly derived from* y, if z can be obtained from y by replacing an occurrence in y of the "left-hand side" of some production by its "right-hand side," e.g., if G has a production $v \to u$, such that y can be written as pvr and z as pur. We write $y \overset{*}{\Rightarrow} z$, and say y *derives* z, or z is *derivable from* y, if $y = z$ or there is some sequence of strings w_1, $w_2 \ldots w_n$, with $w_1 = y$ and $w_n = z$ such that for all i, w_i directly derives w_{i+1}. (Thus $\overset{*}{\Rightarrow}$ is the reflexive transitive closure of $\Rightarrow$.)

Finally, we have the definition of the language generated by a grammar.

3 Definition. Let G be a grammar with start symbol S. The language generated by G, $L(G)$, is defined to be the set of terminal strings derivable from the start symbol:

$$L(G) = \{w | w \in X^* \text{ and } S \overset{*}{\Rightarrow} w\}$$

Given G, if $S \overset{*}{\Rightarrow} w$ (where w is in general made up of both terminal and nonterminal symbols) we refer to w as a *sentential form*. Thus $L(G)$ is the set of terminal sentential forms.

4 Example. Let $V = \{S\}$, $X = \{a, b\}$, $P = \{S \to aSb, S \to ab\}$. In this case, $L(G) = \{a^n b^n | n > 0\}$. That is, $L(G)$ is the set of all strings of the following form: a nonempty block of a's, followed by a block of b's of the same length.

In order to make our descriptions of grammars more concise, we often simply list the productions if it is clear what choice of start symbol, and of terminal and nonterminal symbols, is appropriate. Moreover, we combine productions which have the same left-hand sides using the symbol |. For example, we write the grammar of Example 4 as $S \to aSb | ab$.

5 Example. The grammar

$$S \to aSa | bSb | a | b | \lambda$$

generates the set of all palindromes over the alphabet $\{a, b\}$. A *palindrome* is a string that is the same forward and backward. The set of all palindromes over the terminal alphabet $X = \{a, \ldots, z\}$ includes a number of remarkable linguistic curiosities (blanks, punctuation, and capitals ignored), including

Madam, I'm Adam

A man a plan a canal Panama

Sit on a potato pan Otis

Doc note I dissent a fast never prevents a fatness: I diet on cod

Live dirt up a side track carted is a putrid evil

saippuakauppias (the Finnish word for "soap salesman").

6 Example. The following grammar generates all strings over the terminal alphabet 0, 1 with an equal count of 0's and 1's.

$$V = \{S, A, B\}$$
$$X = \{0, 1\}$$

P comprises the following productions:

$$S \to 0B | 1A$$
$$A \to 0 | 0S | 1AA$$
$$B \to 1 | 1S | 0BB$$

Note that this is our first example of a grammar with other nonterminals besides the start symbol S. How can we prove that this grammar does in fact generate the intended language? This problem—proving the correctness of a grammar—has no general algorithmic solution (in a sense to be made precise in Chapter 5). However, using mathematical induction, it is possible in many important cases to give a proof that a particular grammar does indeed generate a particular language. We now give such a proof for the grammar and language of Example 6.

PROOF. There are really two theorems to prove: we must show first that the grammar in question generates *only* strings with equal numbers of 0's and 1's. Then we must show that *all* strings with equal numbers of 0's and 1's are grammatically derivable.

To prove the first theorem we strengthen the claim slightly. We show that

(1) only equal count strings are derivable from S;
(2) any string derivable from A has an excess of one 0; and
(3) any string derivable from B has an excess of one 1.

The claims follow trivially for strings in the language of length 1 or 2. So suppose the claims hold for every string of length less than n, where n is at least 2. Let w be a string of length n. If w is derivable from S, then its derivation must have begun either with the $0B$ or $1A$ production. Without loss of generality consider the first case. Notice first that the B in $0B$ derives a string of length less than n. Therefore, by the inductive hypothesis, it derives a substring with one extra 1. The 0 at the front of this production then balances this, so that w has as an equal number of 0's and 1's. In a similar fashion we can establish that for strings of length n, A derives strings with an excess 0 ($A \to 0S$ gives an excess 0; $A \to 1AA$ gives a 1 and two excess 0's); and B derives strings with an excess 1. Thus we have established the first part of our theorem.

The proof that any string with an equal number of 0's and 1's is derivable in this grammar—our second theorem—is similar. Again we argue by induction, strengthening our claim by showing that any string with an excess of one 0 is derivable from A, and any string with an excess of one 1 is derivable from B. The basis step is again trivial: any string of length 1 or 2 with an equal count of 0's and 1's is derivable from S; moreover, A derives any string of length 1 or 2 with an excess 0 (namely, 0), and B derives any string of length 1 or 2 with an excess of 1 (namely, 1).

Suppose our various inductive hypotheses hold for all strings of length less than n, and suppose w is of length n. We prove that $S \to w$ if w has an equal number of 0's and 1's.

Suppose without loss of generality that w begins with a 0. Then the rule $S \to 0B$ is taken as the first step in the derivation of w. Now w with its initial 0 removed is of length $n - 1$, and has a single excess 1. Therefore, the inductive hypothesis applies to this string and we can conclude that it is derivable from B. Hence the grammar does indeed derive w. To complete the proof, the reader may use a similar argument to verify that $A \rightarrow w$ if w has an extra 0, while $B \rightarrow w$ if w has an extra 1. □

Incidentally, this proof demonstrates that for grammars, as well as for programs, it is valuable to provide comments—in this case, comments concerning the language derivable from each nonterminal.

7 Example. Consider the following grammar:

$$S \rightarrow aSBC | aBC$$

$$CB \rightarrow BC$$

$$aB \rightarrow ab$$

$$bB \rightarrow bb$$

$$bC \rightarrow bc$$

$$cC \rightarrow cc$$

What strings are produced by this grammar G? Note the following (where, e.g., a^{n-1} is shorthand for a string of $(n-1)$ a's):

using the first production n times we get $S \overset{*}{\Rightarrow} a^{n-1}S(BC)^{n-1}$
using the second production once, we get $a^{n-1}S(BC)^{n-1} \Rightarrow a^n(BC)^n$
using the third production $(n-1)$ times, we get $a^n(BC)^n \overset{*}{\Rightarrow} a^nB^nC^n$
using the fourth production, we get $a^nB^nC^n \Rightarrow a^nbB^{n-1}C^n$
using the fifth production $(n-1)$ times, we get a $a^nbB^{n-1}C^n \overset{*}{\Rightarrow} a^nb^nC^n$
using the sixth production once, and the seventh production $(n-1)$ times, we get $a^nb^nC^n \overset{*}{\Rightarrow} a^nb^nc^n$.

Hence, for all $n > 0$, $a^nb^nc^n \in L(G)$.
One can show that, in fact, $L(G)$ contains no other words.

Notice that whereas the language $\{a^nb^nc^n | n > 0\}$ is similar in an obvious way to the language of Example 2, the generating grammars for the two languages are quite different. In particular, the grammar in this example involves productions with left-hand sides of length greater than one. A *context-free grammar* is a grammar in which the left-hand side of every production is a single nonterminal. Does the language of this example have a generating context-free grammar? We give a negative answer to this question in the next section, where we look more closely at the structure of context-free languages.

EXERCISES FOR SECTION 1.2

1. Write a grammar for the language $\{a^nb^{2n} | n > 0\}$.

2. What language does the grammar $S \rightarrow aSb | SS | ab | ba$ generate? Prove your result.

3. Prove that the grammar of Example 7 generates only strings of the form $a^n b^n c^n$.

4. What language does the grammar $S \to SaSbS|\lambda$ generate? Prove your result.

5. What language does $S \to SaSbS|SbSaS|\lambda$ generate? Prove your result.

6. Consider the following grammar: $S \to SaaSbbS|SbbSaaS|SabSbaS|\lambda$. What production rules must be added so that the resulting grammar generates the set of all strings with an equal number of a's and b's?

7. What language does the grammar $S \to S(S)S|S[S]S|\lambda$ generate?

1.3 Context-Free and Context-Sensitive Languages

In this section we take a first look at two important language classes, the context-free languages and the context-sensitive languages. A grammar is context-free if, unlike the grammar of Example 7 in the last section, the replacement of a nonterminal is not restricted by the context (adjacent letters) in which it occurs.

1 Definition. A grammar G is *context-free* if v is a single nonterminal for every production $v \to w$ in P. A language L over some alphabet X is context-free if it can be generated by a context-free grammar.

Thus the language of palindromes, the language of matched parentheses, and the language made up of strings with equal numbers of a's and b's are all context-free, because in each case we exhibited a generating context-free grammar. Note that if $L = L(G)$ for a grammar G which is *not* context-free, it remains to be determined whether or not L is context-free, for there may or may not also be a context-free grammar G' for which $L = L(G')$. As we pointed out earlier, the grammar given for the language $\{w|w$ is of the form $a^n b^n c^n$ for some $n > 0\}$ is not a context-free grammar, and we shall show shortly that *no* context-free grammar exists for this language.

As a generalization of context-free rules, we introduce context-sensitive rules which also specify the replacement of a single nonterminal symbol, but in general also require a context for application. Thus the rule $bC \to bc$ of Example 1.2.7 is context-sensitive because it says that the nonterminal C may be replaced by the terminal symbol c only in the context of a preceding b. More formally we have the following definition:

2 Definition. A grammar G is *context-sensitive* if each production is either of the form

(1) $yAz \to ywz$, for A in V, y, z in $(V \cup X)^*$, w in $(V \cup X)^+$ (i.e., A can be replaced by w in the context $y - z$); or of the form

(2) $S \to \lambda$, provided S does not appear on the right-hand side of any production.

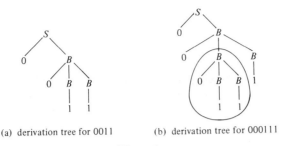

(a) derivation tree for 0011   (b) derivation tree for 000111

Figure 1

A language is *context-sensitive* if it can be generated by a context-sensitive grammar.

Strictly speaking, not every context-free grammar is context-sensitive, since Definition 1 allows productions like $A \to \lambda$. However, we shall soon show how to modify context-free grammars so that every context-free language is a context-sensitive language. Notice also that, because of the production $CB \to BC$, the grammar of Example 1.2.7 is not context-sensitive according to the definition given previously. However, we shall shortly prove (as Theorem 2.1.6) the following result, which establishes that the language of 1.2.7 is in fact context-sensitive.

3 Fact. *Let G be a grammar such that every production of G is of the form $v \to w$, with $|v| \leq |w|$, except that there may be a single λ-production $S \to \lambda$ if S does not appear on the right-hand side of any production. Then there is a context-sensitive grammar G' which is equivalent to G: $L(G) = L(G')$.*

Context-free derivations have a very elegant and useful tree representation. For example, the derivations of the strings 0011 and 000111 using the grammar of Example 1.2.6 are given in Figure 1.

If a string can be derived legally by a *context-free* grammar, then we can describe that derivation by a tree T with the following properties:

(1) The root is labeled with the start symbol S;
(2) Every node that is not a leaf is labeled with a variable—a symbol from V;
(3) Every node that is a leaf is labeled with a terminal—a symbol from X (or possibly with λ);
(4) If node N is labeled with an A, and N has k direct descendants $N_1, \ldots, N_k$, labeled with symbols $A_1, \ldots, A_k$, respectively, then there is a production of the grammar of the form $A \to A_1 A_2 \ldots A_k$;
(5) An expression derived by some derivation may be obtained by reading the leaves of the tree associated with that derivation left to right.

Notice that a tree representation does not order the application of productions in a derivation. Thus Figure 1(a) represents either

$$S \Rightarrow 0B \Rightarrow 00BB \Rightarrow 001B \Rightarrow 0011$$

$$S \Rightarrow 0B \Rightarrow 00BB \Rightarrow 00B1 \Rightarrow 0011$$

The first derivation is referred to as a leftmost derivation, since the leftmost nonterminal in each sentential form is always expanded first. Analogously, the second derivation is called a *rightmost derivation*. The reader will note that the tree of Figure 1(b) corresponds to many derivations besides the leftmost and rightmost.

Given a tree for a context-free derivation, we define the *length of a path* from the root to a leaf to be the number of nonterminals on that path. The *height* of a tree is the length of its longest path. (Thus in Figure 1(a), the height of the tree is 3.)

Consider the circled subtree in Figure 1(b). It is a legal B-tree; i.e., it is a legal derivation tree using the grammar from Example 1.2.6 except that its root is a B and not an S. Now if we take any legal derivation tree for a string in this language and replace any B-subtree in it with the circled subtree, we obtain another legal derivation tree. This is the meaning of "context-free" from the point of view of tree representations. We next show how systematic application of this subtree replacement principle can be used to establish a key result about the structure of context-free languages.

Suppose we are given a derivation tree T for a string z of terminals generated by some grammar G, and suppose further that the nonterminal symbol A appears twice on some path, as shown in Figure 2, where $z = uvwxy$. Here the lower A-tree derives the terminal string w, and the upper A-tree derives the string vwx. Since the grammar is context-free, replacing the upper A-tree with the lower A-tree does not affect the legality of the derivation. The new tree derives the string uwy.

On the other hand, if we replace the lower A-tree with the upper A-tree we obtain a legal derivation tree for the string $uvvwxxy$, which we can write as uv^2wx^2y. This upper-for-lower substitution may be repeated any finite number of times, yielding the set of strings $\{uv^nwx^ny \mid n \geq 0\}$. Every infinite context-free language must contain infinite subsets of strings of this general form. In a

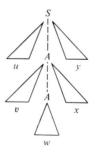

Figure 2   A derivation tree with the nonterminal symbol A appearing twice on the same path.

moment we will use this important fact to show that the language of Example 1.2.7 has no context-free grammar. But first we need to examine our result about subtree substitutions in more detail. We begin with a lemma that relates the length of a string to the depth of a derivation tree for the string.

4 Lemma. *Let G be a context-free grammar, and suppose the longest right-hand side of any production has m symbols. Let T be a derivation tree for a string w in L(G). If no path of T is of length greater than j, then* $|w| \leq m^j$.

PROOF. As is often the case with inductive proofs, we actually establish a slightly stronger result: if T is a derivation tree for a string w which is rooted at any symbol (not necessarily the start symbol) and T has no path of length greater than j, then $|w| \leq m^j$. We proceed by induction on j. As base case, suppose $j = 1$. Then T can represent only a single production, so that only a string of length m or less can be produced.

So suppose the result holds for trees T, rooted at any nonterminal symbol, which have height $\leq j$. Let T be rooted at A, and suppose T is of height $j + 1$. Suppose the tree top comes from the production $A \rightarrow v$. Then each symbol in v can, at most, command a tree of height j. Therefore, by induction, each of these trees adds at most a length of m^j to the length of w (v might also contain terminals, and these only add one symbol). Since $|v| \leq m$, $|w| \leq m \cdot m^j = m^{j+1}$. $\square$

5 The Pumping Lemma (Also Known as the *uvwxy* Theorem). *Let L be a context-free language. Then there exist constants p and q depending on L only, such that if z is in L with* $|z| > p$, *then z may be written as uvwxy in such a way that*

(1) $|vwx| \leq q$;
(2) *at most one of v and x is empty; and*
(3) *for all* $i \geq 0$, *the strings* uv^iwx^iy *are in L.*

PROOF. Let G be a context-free grammar generating L for which m is the length of the longest right-hand side of any production. Suppose that for this G, V has k elements. We now set $p = m^k$. Then by the lemma, if $|z| > p$, then in any derivation tree for z the length of some path must be at least $k + 1$.

Since G has only k variables, some variable must be duplicated on that path (see Figure 2). Consider a lowest duplicated pair, and call the duplicated variable A. Then the upper A of this pair contains no other duplicates below it. This means that the path from the upper A to the leaf is of length at most $k + 1$. Call the string derived from the subtree rooted at the upper A vwx, where w is the string derived by the lower A-subtree. The length of vwx is at most $m^{k+1} = q$, by the lemma.

Let us write z as $uvwxy$, and apply the tree substitution principle described earlier. If we substitute the lower A-subtree for the upper A-subtree, the derived string is $uwx = uv^0wx^0y$. If we substitute the upper tree for the lower tree i times, we get $uv^{i+1}wx^{i+1}y$.

We conclude the proof by showing that v and x cannot both be empty for every choice of a path with a duplicated pair. For if $v = x = \lambda$, we can substitute the lower A-subtree for the upper A-subtree without changing the string z derived by the tree as a whole. If all paths of length at least $k + 1$ could be shortened in this way without changing the derived string, we would obtain a derivation tree for z with height less than $k + 1$, which is impossible.   □

Note the following:

(1) Given a language generated by a grammar that is not context-free you cannot deduce immediately that it is not also generated by a context-free grammar. Just being unable to think of such a grammar is no guarantee that none exists.

(2) But if an infinite language does not obey the pumping lemma, it cannot be generated by a context-free grammar.

6 Example. The language $\{a^nb^nc^n|n > 0\}$ is not context-free.

PROOF. Suppose, by way of contradiction, that this language is context-free, so that there exist p and q as in the statement of pumping lemma. Consider the string $a^rb^rc^r$, with $r > p$ and $r > q$. Then the pumping lemma applies to this string, which we can write as $z = uvwxy$. The vwx substring could fall (1) entirely inside the a's block, b's block, or c's block; or (2) it could fall across a single boundary (the $a - b$ boundary or the $b - c$ boundary). It cannot fall across two boundaries, however, because it is not long enough, given our choice of r. Hence for any of these cases, consider the pumped string $uvvwxxy$. In this string, the count of at least one of the symbols a, b, or c must not increase, and so a string not of the form $a^nb^nc^n$ is obtained. This is a contradiction.   □

7 Example. The language $\{a^k|k$ is a perfect square$\}$ is also not context-free. This result also follows by contradiction using the pumping lemma. Suppose the language is context-free; there again exist appropriate p and q. We can choose $p \geq q$ without denying the validity of the pumping lemma. Set $z = a^{p^2}$. Since $|z| > p$, we can write $z = uvwxy$ with $|vwx| < q \leq p$. But then $|uvvwxxy| \leq |z| + |vwx| = p^2 + q < p^2 + 2p + 1 = (p + 1)^2$. Hence $|uvvwxxy|$ is *not* a perfect square.

EXERCISES FOR SECTION 1.3

1. Write grammars for the following languages.
   (a) Arithmetic expressions involving the digits 0 and 1 and the operations $+$ and $*$. Example: $(1 + (0 * 1))$
   (b) $\{w|w$ is of the form a^nb^m, with $n < m\}$
   (c) Formulas of the propositional calculus with two variables, p and q, and connectives and, or, and not. Example: $(p$ or $(q$ and $($not $p)))$

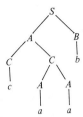

Figure 3   A derivation tree for a grammar G.

(d) $\{ww^R|w$ in $\{a,b\}^*\}$ where w^R stands for the reversal of w, e.g., if $w = abaa$, then $w^R = aaba$.

2. Consider the derivation tree given in Figure 3 for some grammar G. Describe (briefly!) one million strings in $L(G)$.

3. Write a grammar for the language $\{a^k|k$ is a perfect square$\}$. (*Hint*: Use the fact that $1 + 3 + 5 + \cdots + (2n - 1) = n^2$.)

4. Use the pumping lemma to prove that the following languages are not context-free.
   (a) $\{a^i b^{2i} c^{3i}|i > 0\}$
   (b) $\{a^i|i$ is a prime$\}$
   (c) $\{a^i b^j|i = j^2\}$
   (d) $\{ww|w$ in $\{a,b\}^*\}$

5. Suppose a grammar G has trivial productions, i.e., productions of the form $A_1 \to A_2$, $A_2 \to A_3, \ldots, A_n \to A_1$. Show that $L(G)$ can be generated by a grammar G' in which all such cycles have been removed. (*Hint*: You may want to add new productions.)

6. In unary notation an integer $n \geq 0$ is represented by a string of $n + 1$ 1's. Thus "111" $= 2$, "1" $= 0$, etc.
   (a) Write a grammar for all equations (not necessarily correct!) that can be formed from fully parenthesized expressions involving integers in unary notation, $=, +,$ and $*$, e.g., $((1 + 111) * 111) = (111111111 + (1111111 * 1111111))$.
   (b) Show that the set of *correct* equations from the language of part (a) (e.g., $((11 * 1111) = 1111))$ is *not* context-free.

7. Write a grammar for the language $\{a^n b^n c^n d^n|n > 0\}$.

8. Is the language $\{a^n b^k c^n d^k|n, k > 0\}$ context-free?

9. Use the pumping lemma to show that the language $\{a^i b^j c^k|0 < i \leq j \leq k\}$ is not context-free. (*Hint*: It may be helpful to use the pumping lemma both forward and backward.)

1.4 Programs, Languages, and Parsing

As we pointed out at the beginning of this chapter, context-free languages are important for computer science because they are a reasonably adequate mechanism for specifying the syntax of programming languages. In this section we take a first look at this specification mechanism.

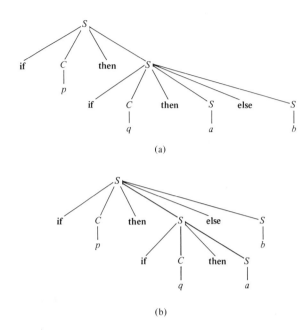

Figure 4   Two different derivation trees for **if** p **then if** q **then** a **else** b.

We begin with a simple example, the **if-then** and **if-then-else** programming constructs which are present in many programming languages. As a first approximation, consider the following grammar:

$$S \rightarrow \textbf{if } C \textbf{ then } S \textbf{ else } S | \textbf{if } C \textbf{ then } S | a | b$$

$$C \rightarrow p | q$$

Here S is a (statement) nonterminal, C is a (conditional) nonterminal, a and b are (statement) terminals, p and q are (condition) terminals, and **if, then,** and **else** are (reserved word) terminals.

There are problems with this grammar. It generates the intended language, but it does so ambiguously. In particular,

if p **then if** q **then** a **else** b

can be generated in two ways—as in either Figure 4a or Figure 4b—corresponding to two different interpretations of the statement:

if p **then (if** q **then** a **else** b)

and

if p **then (if** q **then** a) **else** b.

From a programming point of view, one standard way of interpreting such constructions is to associate each **else** statement with the nearest **if**. If a grammar is to reflect this agreed programming preference, then one form—in this case the first one—must be generable by the grammar, and the alternative must be excluded. We give such a grammar in the following example.

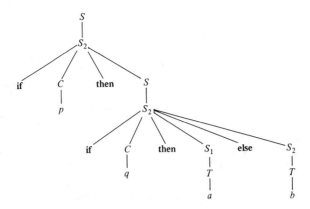

Figure 5   The unique derivation tree of **if p then if** q **then** a **else** b using the grammar of Example 1.

1 Example (An Alternative **if-then-else** Grammar)

$$S \to S_1 | S_2$$
$$S_1 \to \textbf{if } C \textbf{ then } S_1 \textbf{ else } S_2 | T$$
$$S_2 \to \textbf{if } C \textbf{ then } S | \textbf{if } C \textbf{ then } S_1 \textbf{ else } S_2 | T$$
$$C \to p | q$$
$$T \to a | b$$

This grammar generates only one possible derivation of the string

if p **then if** q **then** a **else** b

namely, the derivation in which the final **else** statement, b, is linked to the nearest **if**, as shown in Figure 5.

Grammars that are unambiguous are clearly essential for well-defined syntactic specification, for otherwise a compiler could translate a program in different ways to direct completely different computations. As a result, the study of ambiguity is an important practical as well as theoretical aspect of formal language theory.

2 Definition. A context-free grammar G is *ambiguous* if some string x in $L(G)$ has two distinct derivation trees. Alternatively, G is ambiguous if some string in $L(G)$ has two distinct leftmost (or rightmost) derivations. A grammar is *unambiguous* if it is not ambiguous, and a language L is *inherently ambiguous* if *every* grammar for L is ambiguous.

The **if-then-else** example just considered illustrates how grammars can be used to generate programming language constructions. Context-free descriptions must also have a second property if they are to be a useful programming language formalism. Given a string (a program text) and a (programming) language specified by some grammar, it must be possible to construct a derivation tree for the string quickly and easily if the string belongs to the language, or report an error if the string is not well formed. The problem of doing string derivation backward—given a string, recover its derivation—is known as the *parsing problem* for context-free languages. Its satisfactory solution is one of the most important milestones in the development of computer science.

For the present, let us concern ourselves with a somewhat simplified view of parsing: given a string x, how can we tell if x is legally generable by a particular context-free grammar G? There are two basic strategies for solving this problem. One strategy, "bottom-up" parsing, attempts to construct a parse tree for x by hypothesizing a derivation tree for x beginning with the bottom—the tips of the tree—and working up the tree to the root. The second basic strategy, "top-down" parsing, works the other way around, hypothesizing the top of the derivation tree first beginning with the root. We illustrate these two techniques in the next two examples.

3 Example. Parsing arithmetic expressions bottom to top. In this example we give a grammar for a fragment of the language of arithmetic expressions and explain a simple algorithm for parsing these expressions bottom-up. The start symbol for the grammar is E.

$$(1) \ E \to E + T$$

$$(2) \ E \to T$$

$$(3) \ T \to T * F$$

$$(4) \ T \to F$$

$$(5) \ F \to (E)$$

$$(6) \ F \to 2$$

Now suppose we wish to decide whether the string

$$2 + 2 * 2$$

is generated by this grammar. Our approach is to run the derivation backward, mimicking a rightmost derivation as we process the string from left to right. The first five steps in the process look like this:

$$2 + 2 * 2 \Leftarrow F + 2 * 2 \ (\text{reverse } 6)$$

$$\Leftarrow T + 2 * 2 \ (\text{reverse } 4)$$

$$\Leftarrow E + 2 * 2 \ (\text{reverse } 2)$$

$$\Leftarrow E + F * 2 \ (\text{reverse } 6)$$

$$\Leftarrow E + T * 2 \ (\text{reverse } 4)$$

And now we are at a crucial point in the parse. Three reversals are possible: we could

(1) convert $E + T$ to E, leaving $E * 2$;
(2) convert T to E, leaving $E + E * 2$;
(3) convert 2 to F, leaving $E + T * F$.

You should convince yourself that the first two choices are not viable—they lead to "dead ends" from which no successful parse is possible. Choice (3) alone leads to a successful parse, by

(1) converting $T * F$ to T; and
(2) converting $E + T$ to E, the grammar's start symbol.

This example illustrates the nondeterminism possible in the parsing process. The design of efficient parsers—parsers that limit, or eliminate completely, the kind of nondeterminism we have seen in this example—is a major component of the area of computer science known as syntactic analysis. We shall examine syntactic analysis in more detail in Chapter 7.

4 Example. Our next parsing example illustrates the second general parsing principle: top-down parsing. We illustrate this principle by showing how to parse a PASCAL fragment called the language of **while**-programs. This simple language permits successor, predecessor, and zero assignment statements, as well as while loops with simple tests of the form $x \neq y$. In the companion volume *A Programming Approach to Computability* [Kfoury, Moll, Arbib, 1982] in this series, we study the syntax and semantics of this language in detail. A typical program in this language is given in the following:

$$
\begin{array}{ll}
\textbf{begin} & y := 0; \\
& \textbf{while } x \neq y \textbf{ do} \\
& y := \textbf{succ}(y) \\
\textbf{end} &
\end{array}
$$

This program sets the value of y equal to that of x, i.e., the program performs the assignment statement $y := x$.

We now give a simple grammar for the syntax of this language. For this grammar,

$$V = \{C, S, S_1, S_2, A, W, C, U, T\}$$
$$X = \{\textbf{begin}, \textbf{end}, \textbf{pred}, \textbf{succ}, :=, \neq, \textbf{while}, \textbf{do}, :, (,), 0, x, y\}$$
Start Symbol $= C$
Productions $=$
$\quad C \rightarrow \textbf{begin } S_1 \textbf{ end } \{C \text{ for compound statement}\}$
$\quad S_1 \rightarrow S \, S_2$
$\quad S_2 \rightarrow ; S_1 | \lambda$
$\quad S \rightarrow A | W | C$
$\quad A \rightarrow V := T \{A \text{ for assignment statement}\}$
$\quad T \rightarrow \textbf{pred}(V) | \textbf{succ}(V) | 0$
$\quad W \rightarrow \textbf{while } V \neq V \textbf{ do } S \{W \text{ for while statement}\}$
$\quad V \rightarrow x | y$

We have simplified the language of **while**-programs for the purposes of this example by assuming that only two variables, x and y, are present in the language.

Let us now attempt to parse the simple program given earlier in top-down fashion. First of all, since C is the start symbol of the grammar, any legal parse must begin

$$\textbf{begin } S_1 \textbf{ end}$$

because this is the only C production. Next we examine the program text and notice that the **begin**'s match, so we continue the parse. Our strategy at each stage is to attempt to expand the leftmost variable in the current string. The next input symbol is y. There is only one S_1 production, $S_1 \rightarrow S\ S_2$, and using this production it is possible to reach the symbol y. So we try this S_1 expansion, arriving at

$$\textbf{begin } S\ S_2 \textbf{ end.}$$

Next we attempt S expansion. There are three S rules, but only $S \rightarrow A$ is applicable, since A can reach a leading variable, but W leads to **while** ..., and C leads to **begin** .... So we obtain

$$\textbf{begin } A\ S_2 \textbf{ end.}$$

There is only one A production, so we apply it to get

$$\textbf{begin } V := T\ S_2 \textbf{ end.}$$

V can go to either x or y, and since the next input symbol is y we make that conversion. Our new form looks like

$$\textbf{begin } y := T\ S_2 \textbf{ end.}$$

We have now matched the input up to :=, and we attempt to expand the T on the next input symbol, 0. There are three T productions, but only the $T \rightarrow 0$ is relevant, so we get

$$\textbf{begin } y := 0\ S_2 \textbf{ end.}$$

The new current input symbol is ;, and this matches the S_2 production $S_2 \rightarrow ; S_1$, and fails to match $S_2 \rightarrow \lambda$, and so we obtain

$$\textbf{begin } y := 0; S_1 \textbf{ end.}$$

As before, $S_1 \rightarrow S\ S_2$, and in this case, $S \rightarrow W$ is the only relevant production:

$$\textbf{begin } y := 0; W\ S_2 \textbf{ end.}$$

W generates the **while** statement as its only right-hand side, yielding

$$\textbf{begin } y := 0; \textbf{ while } V \neq V \textbf{ do } S\ S_2 \textbf{ end.}$$

The two nonterminal V's match the x and y, respectively, from the text, and we get

$$\textbf{begin } y := 0; \textbf{ while } x \neq y \textbf{ do } S \ S_2 \textbf{ end.}$$

The next nonterminal is S, and it must be replaced by A, since the next text item is a variable:

$$\textbf{begin } y := 0; \textbf{ while } x \neq y \textbf{ do } A \ S_2 \textbf{ end.}$$

Proceeding exactly as we did with our earlier assignment statement, we can replace A by the appropriate expression to obtain

$$\textbf{begin } y := 0; \textbf{ while } x \neq y \textbf{ do } y := \textbf{succ}(y) \ S_2 \textbf{ end.}$$

Finally we convert S_2 to the empty string, λ, since no; is forthcoming. The **end**'s match, and so we have successfully parsed the program. Notice that our method was completely deterministic: no backtracking was required, and at every stage we could predict with no difficulty which production was appropriate, given the current text symbol. This style of parsing is intuitively pleasing and easy to program.

The top-down technique just developed is called LL parsing. In Chapter 7 we discuss (strong) LL parsing in detail, and we identify the grammar we have just considered as a strong $LL(1)$ grammar.

EXERCISES FOR SECTION 1.4

1. Describe a bottom-up parse of the string $2 * (2 + 2)$ using the grammar of Example 3.

2. Describe a top-down parse of the program text

$$\textbf{begin while } x \neq y \textbf{ do}$$
$$\textbf{begin } x := 0;$$
$$y := 0$$
$$\textbf{end}$$
$$\textbf{end}$$

using the grammar of Example 4.

3. Suppose G is a context-free grammar from which the string 010110 can be derived. Suppose the derivation tree for 010110, given in Figure 6, includes all of the productions of G. Show that G is ambiguous.

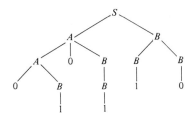

Figure 6   A derivation tree using all productions of a grammar G.

1.5 Context-Free Grammars and Natural Language

At the beginning of this chapter we pointed out that the modern theory of formal grammars stems in part from theories of natural language that were developed by the linguist Noam Chomsky. In this section we briefly describe how formal grammars can be used to describe natural language.

Suppose we identify the following language categories, which we associate with nonterminal symbols of a grammar:

S—Sentence
NP—Noun phrase
VP—Verb phrase
Adj—Adjective
Det—Determiner
V—Verb
N—Noun
$Prep$—Preposition
A—Adjectival phrase
PP—Prepositional phrase

Given these symbols, we can write a context-free grammar which (roughly) describes the syntactic structure of a small subset of English.

$S \rightarrow NP\ VP$
$NP \rightarrow Det\ N | Det\ A\ N$
$A \rightarrow Adj\ A | Adj$
$VP \rightarrow V_{be}\ PP$
$PP \rightarrow Prep\ NP$

Of course, the words of English constitute the terminal symbols of our grammar, and so we must add productions of the form

$V_{be} \rightarrow is | are | was \ldots$
$Det \rightarrow the | an | a \ldots$
$Prep \rightarrow on | in | for \ldots$

etc. Given this grammar we can derive sentences like "The big red ball is on the table" as follows:

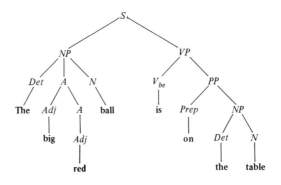

One immediate problem with this grammar is case agreement. The grammar generates such nonsentences as "The big red ball are on the table," and so if our grammar is to be a more adequate description of English syntax, machinery which ensures subject verb agreement must be introduced.

A more serious objection to the system described previously can be seen by considering sentences like "What is the ball on." This sentence has the following (illegal) structure

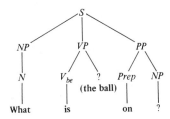

Now this sentence is, in a certain sense, a distortion of the sentence "The ball is on what?". A large body of linguistic theory is concerned with characterizing the ways in which such apparently non-context-free distortions can happen.

Recently linguists have reconsidered the idea of giving a context-free description of English that accounts for many of the anomalies (e.g., a prepositional phrase with empty object) that clearly form an integral part of natural language. We shall develop one such description in Chapter 9.

EXERCISES FOR SECTION 1.5

1. Draw a parse tree for the sentence "The ball is on the big big table."

2. Using the top-down algorithm of the last section, describe how you might parse the sentence in Exercise 1.

CHAPTER 2

Grammars and Machines

In Chapter 1 we had a first look at context-free grammars and context-free languages, and we had a glimpse at two areas of practical interest—parsing and natural language processing—where formal grammatical descriptions play an important role.

In this chapter we will take a broader look at grammars. In particular, we will look at the *Chomsky hierarchy*, a scheme for classifying grammars based on the form of the grammar production rules. We will also examine in depth the *regular languages*, which constitute a proper subset of the context-free languages.

2.1 The Chomsky Hierarchy

In the last chapter we had our first view of context-free grammars, which permit production rules of the form $A \rightarrow w$. These grammars are a special case of the general phrase structure grammars of Definition 1.2.1. In the unrestricted rewriting systems of Definition 1.2.1, any production of the form $v \rightarrow w$ is permitted with v a string of $(V \cup X)^+$ containing at least one nonterminal, and w an arbitrary string in $(V \cup X)^*$.

As a *generalization* of context-free rules, we now introduce context-sensitive rules which also specify the replacement of a single nonterminal symbol, but in general require a context for application. Thus the rule $bC \rightarrow bc$ of Example 1.2.7 is context-sensitive because it says that the non-terminal C may be replaced by the terminal symbol c only in the context of a preceding b. More formally we have the following definition:

1 Definition. A grammar G is *context-sensitive* if each production is either of the form

(1) $yAz \to ywz$, for A in V, y, z in $(V \cup X)^*$, w in $(V \cup X)^+$ (i.e., A can be replaced by w in the context $y - z$); or of the form
(2) $S \to \lambda$, provided S does not appear on the right-hand side of any production.

A language is *context-sensitive* if it can be generated by a context-sensitive grammar.

Strictly speaking, not every context-free grammar is context-sensitive, because context-free productions may be of the form $A \to \lambda$. However, we shall soon see that every context-free language is indeed a context-sensitive language.

In this chapter, we will also study a very useful subclass of the context-free languages, the right-linear languages.

2 Definition. A grammar $G = (V, X, S, P)$ is said to be *right-linear* if every production is of the form

$$A \to bC \quad \text{or} \quad A \to b$$

where A and C are in V and $b \in X \cup \{\lambda\}$. A language generated by such a grammar is called a *right-linear language*.

3 Example. Consider the right-linear language G_1 with productions

$$S \to \lambda |0|0S|1|1S$$

Clearly $L(G_1) = \{0, 1\}^*$, the set of all binary strings.
Now consider the grammar G_2 with productions

$$S \to \lambda |0S|1T$$

$$T \to 0T|1S$$

You should be able to convince yourself that

$$L(G_2) = \{w | w \in \{0, 1\}^* \text{ and } w \text{ has an even number of 1's}\}$$

It will turn out that the four language classes we have identified so far form a *hierarchy*: the right-linear languages are a proper subset of the context-free languages, which are a proper subset of the context-sensitive languages, which in turn are a proper subset of the languages generated by unrestricted rewriting systems. Before stating this result formally, however, we consider a technical lemma which will simplify our analysis of the hierarchy.

4 Lemma (The Empty String Lemma). *Let G be a context-free grammar which involves rules of the form $A \to \lambda$ for arbitrary $A \in V$. Then there exists a* *context-free grammar G' such that*

(i) $L(G) = L(G')$;
(ii) *If $\lambda \notin L(G)$ then there are no productions of the form $A \to \lambda$ in G'; but*
(iii) *If $\lambda \in L(G)$, then there is a single λ-production in G', $S' \to \lambda$, where S' is the start symbol of G' and S' does not appear on the right-hand side of any production in G'.*

PROOF. Suppose $\lambda \notin L(G)$. For $C \in V$ consider all productions

$$C \to w$$

where w is not empty. Now suppose w contains variables $A_1, \ldots, A_k, B_1, \ldots,$ B_j where A_i, $1 \le i \le k$, are precisely those variables in w for which $A_i \overset{*}{\Rightarrow} \lambda$. Add to G the production $C \to w'$, where w' is obtained from w by deleting from w zero or more occurrences of a variable from the A_i's. Thus, for example,

$$C \to aA_1BA_2$$

yields 4 productions

$$C \to aA_1BA_2 | aBA_2 | aA_1B | aB$$

When this process is complete, remove all λ-productions, including any new λ-productions that have been added in the process. The resulting grammar is G'.

Then $L(G) = L(G')$ as follows. If a derivation of some string $w \in L(G)$ involved λ-productions, then the derivation tree for w will have tip nodes labeled with λ, as in the following example.

$S \to ABDC$

$C \to RS$

$B \to \lambda$

$A \to a$

$D \to d$

$R \to \lambda$

$S \to d$

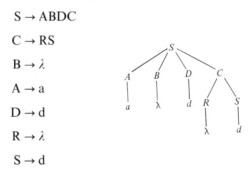

To obtain a G' derivation for w from a G derivation for w, start at the top of the G tree and delete any subtree of the tree root that derives λ. The tree top must now illustrate a production from G'. Then consider the immediate successors of the root node which do not derive λ. Apply the same principle to each of them: if any of their subtrees derive λ, delete that subtree. Continue in this way down the tree. The final tree clearly represents a G' derivation. Furthermore, this tree derives w, because a substring of w is removed by the deletion process if and only if it is λ.

Conversely, if $w \in L(G')$ and the derivation of w involves a newly added rule, then simulate the G' derivation by using the appropriate original G rule and λ-rules. That is, do the process described above in reverse, again proceeding top to bottom.

In the case where $\lambda \in L(G)$, first add a new start symbol to G', say S', and add two new productions, $S' \rightarrow S$ and $S' \rightarrow \lambda$. Next repeat the process described previously on all but the S' productions. Then (see Exercise 1) $L(G) = L(G')$.

$\square$

Using the technique of Lemma 1 it is possible to state and prove an empty string lemma for right-linear grammars, and this is requested in Exercise 2.

The four classes of grammars and corresponding languages introduced previously (right-linear, context-free, context-sensitive, unrestricted) are often referred to by different terminology.

5 Definition. Let $G = (V, X, S, P)$ be a grammar. Then G is classified as type i, $i = 0, 1, 2, 3$, according to the following restrictions on the form of the replacement rules of P:

(i) A grammar is *type 3* if every production is of the form $A \rightarrow bC$ or $A \rightarrow b$, where A, $C \in V$, and $b \in X$ or $b = \lambda$. Such grammars are also referred to as *right-linear grammars*. A language generated by such a grammar is called a type 3 or right-linear language.

(ii) A grammar is *type 2* if every production of P is of the form $A \rightarrow w$, with $A \in V$, and $w \in (V \cup X)^*$. These, of course, are the *context-free grammars*, and the languages they generate are the context-free languages.

(iii) A grammar G is *type 1* if every production of P is of the form $vAw \rightarrow vzw$ where $z \in (V \cup X)^+$ (i.e., $z \neq \lambda$). In addition, we permit the single λ-rule $S \rightarrow \lambda$ in the case where S does not appear on the right-hand side of any production. Type 1 grammars are precisely the *context-sensitive grammars*, and they generate the context-sensitive languages.

(iv) Any grammar G is *type 0*. There are no restrictions on the form of productions for grammars in this class.

Definitions 1 and 2, together with the empty-string lemmas for right-linear and context-free grammars, allow us to conclude that the four classes are contained in one another in ascending order; that is, every type 3 language is also type 2, every type 2 language is also type 1, and every type 1 language is also type 0. We shall prove that these containments are in fact proper, and therefore that the classes form a hierarchy, the so-called Chomsky hierarchy.

We next turn to a normal form result for context-sensitive languages, which will permit us to show that context-free languages are indeed a proper subset of the context-sensitive languages.

6 Theorem. *A language L is context-sensitive if and only if there is some grammar G such that $L = L(G)$ where every production of G of the form $u \rightarrow v$ has the property that $0 < |u| \leq |v|$ with one exception: if $\lambda \in L(G)$, then the rule*

S → λ is also present, and in this case S may not appear on the right-hand side of any production.

PROOF. The "only-if" part of the theorem follows immediately from the definition of a context-sensitive grammar. To show the "if" part, suppose $u \to v$ is a typical production rule with $|u| \leq |v|$, say

$$A_1 A_2 \ldots A_m \to B_1 B_2 \ldots B_n, \qquad m \leq n.$$

The A's and B's may be in either V or X. However, we assume without loss of generality that all the A's are variables. For suppose A_j is a terminal symbol; then we can replace it in the production with a new variable not in V, A'_j, and we add the further production $A'_j \to A_j$.

Let $C_1 \ldots C_m$ be new variables not present in G. We use the C's to build new, context-preserving rules that will take the string $A_1 A_2 \ldots A_m$ and eventually produce $B_1 B_2 \ldots B_n$. Here are the new rules:

$$A_1 A_2 \ldots A_m \to C_1 A_2 \ldots A_m$$

$$C_1 A_2 \ldots A_m \to C_1 C_2 A_3 \ldots A_m$$

$$\ldots$$

$$C_1 C_2 \ldots C_{m-1} A_m \to C_1 C_2 \ldots C_{m-1} C_m B_{m+1} \ldots B_n$$

$$C_1 \ldots C_{m-1} C_m B_{m+1} \ldots B_n \to C_1 \ldots C_{m-1} B_m B_{m+1} \ldots B_n$$

$$\ldots$$

$$C_1 B_2 \ldots B_n \to B_1 B_2 \ldots B_n$$

We leave it as an exercise to show that if G is a grammar whose production rules satisfy the hypotheses of the theorem, then the context-sensitive grammar G' constructed according to the procedure described earlier will yield a language $L(G')$ for which $L(G) = L(G')$. □

Consider again the grammar of Example 1.2.7. Because of the production $CB \to BC$, this grammar is not context-sensitive according to Definition 1. However, the language it generates, $\{a^n b^n c^n | n > 0\}$, is indeed context-sensitive according to Theorem 6.

Let us use the notation $\mathcal{L}_i$ for the Type i languages:

$$\mathcal{L}_i = \{L \subset X^* | L = L(G) \text{ for some Type } i \text{ grammar } G\}.$$

7 The Hierarchy Theorem. *The Chomsky hierarchy is a strict hierarchy of language classes:*

$$\mathcal{L}_3 \subsetneqq \mathcal{L}_2 \subsetneqq \mathcal{L}_1 \subsetneqq \mathcal{L}_0.$$

PROOF. The extension $\mathcal{L}_2 \subset \mathcal{L}_1$ is proper because $\{a^n b^n c^n | n \geq 1\}$ is context-sensitive but not context-free. In Section 3, Example 2.3.20, we shall show that the language $\{a^n b^n | n \geq 1\}$ is context-free but not right-linear. We postpone the proof that Type 1 is properly contained in Type 0 to Chapter 5.   □

In this section we have used the form of grammatical descriptions to give a hierarchical classification of formal languages. In the next section we look at some elementary properties of these language classes.

EXERCISES FOR SECTION 2.1

1. Complete the proof of Lemma 4.

2. State and prove an empty-string lemma for right-linear grammars.

3. Write a context-sensitive grammar for the language $\{a^n b a^n c a^n | n > 0\}$.

4. Write a Type 0 grammar for the perfect squares, expressed in binary, e.g., 1, 100, 1001, etc.

5. In the proof of Lemma 4 we have to consider all the variables A in a context-free grammar which derive the empty string, i.e., for which $A \stackrel{*}{\Rightarrow} \lambda$. Describe carefully an algorithm that will detect all variables A for which $A \stackrel{*}{\Rightarrow} \lambda$. (*Hint:* Proceed inductively. First find all A's for which $A \stackrel{1}{\Rightarrow} \lambda$, then having found those for which $A \stackrel{m}{\Rightarrow} \lambda$, find those for which $A \stackrel{m+1}{\Rightarrow} \lambda$. $A \stackrel{m}{\Rightarrow} \lambda$ means "A derives λ in m steps but not in $m - 1$ steps.")

6. Let $G = (V, X, S, P)$ where $V = \{S\}$, $X = \{a, b\}$, and P is the following set of productions:

$$S \to aSb$$
$$S \to aSa$$
$$S \to bSa$$
$$S \to bSb$$
$$S \to \lambda$$

Show that $L(G)$ is a right-linear language.

7. Suppose that $G = (V, X, S, P)$ is a context-free grammar such that each production in P is either of the form $A \to wB$ or of the form $A \to Bw$ or of the form $A \to w$, where $A, B \in V$ and $w \in X^*$. Is $L(G)$ necessarily a right-linear language? Prove this claim or find a counterexample.

8. Find a right-linear grammar for the language $\{w \in \{a, b\}^* | w$ does not contain the substring $bab\}$.

9. Let $G = (V, X, S, P)$ be a context-free grammar. Describe carefully an algorithm that will decide for arbitrary strings $w, w' \in (V \cup X)^*$ whether $w \stackrel{*}{\underset{G}{\Rightarrow}} w'$. (*Hint:* See Exercise 5.)

10. (a) Let $G = (V, X, S, P)$ be a grammar where every production is of the form

$$A \to wB \quad \text{or} \quad A \to w$$

where $A, B \in V$ and $w \in X^*$. Show that $L(G)$ is right-linear.

(b) A grammar $G = (V, X, S, P)$ is *linear* if every production is of the form

$$A \to wBw' \quad \text{or} \quad A \to w$$

where $A, B \in V$ and $w, w' \in X^*$. Show that the right-linear languages form a proper subclass of the linear languages. (*Hint*: Consider the language $\{a^n b^n | n \geq 1\}$ mentioned in the proof of Theorem 7.)

11. (a) Prove the following pumping lemma for linear languages. If L is a linear language, then there is a constant n such that if $z \in L$ is of length $\geq n$, then $z = uvwxy$ where $|uvxy| \leq n$, $|vx| \geq 1$, and $uv^iwx^iy \in L$ for all $i \geq 0$.
    (b) Show that the language $\{a^i b^i c^j d^j | i \geq 1 \text{ and } j \geq 1\}$ is context-free but not linear.

2.2 Closure Properties

Language theorists have long been interested in ways in which given languages (sets of words) can be combined to form new ones. We start with the three most basic methods studied in language theory:

1 Definition. Let L_1, L_2 be two languages over the same alphabet X (i.e., L_1 and L_2 are both subsets of X^*). Then we define:

(i) The *union* of L_1 and L_2 is (as usual)

$$L_1 \cup L_2 = \{w | w \in L_1 \text{ or } w \in L_2\}$$

(ii) The *concatenation* of L_1 and L_2 (also called "L_1 dot L_2") is

$$L_1 \cdot L_2 = \{w | w = w_1 w_2 \text{ for some } w_1, w_2 \text{ with } w_1 \in L_1, w_2 \in L_2\}$$

(iii) The *iterate* of L_1 (also called "L_1 star") is

$$L_1^* = \{w | w = w_1 w_2 \ldots w_n \text{ for some } n \geq 0 \text{ and each } w_i \in L_1\}$$
$$= \bigcup_{n \geq 0} L_1^n$$

where we define the powers L_1^n of L_1 for $n \geq 0$ by

$$\textit{Basis Step: } L_1^0 = \{\lambda\}$$

$$\textit{Induction Step: } L_1^{n+1} = L_1^n \cdot L_1 \text{ for } n \geq 0$$

Thus $L_1^1 = L_1$, $L_1^2 = L_1 \cdot L_1$, etc.

We say a *class* of languages is closed under a language operation if, whenever we apply that operation to languages which belong to the class, we get another language belonging to the class. More formally:

2 Definition. Fix an alphabet X. Let $\mathscr{L}$ be a class of languages over X (i.e., $\mathscr{L} \in 2^{2^{X^*}}$, or $L \in \mathscr{L}$ implies $L \subset X^*$).

Let α be a binary operator on languages, i.e., $\alpha\colon 2^{X^*} \times 2^{X^*} \to 2^{X^*}, (L_1, L_2) \mapsto \alpha(L_1, L_2)$; and let β be a unary operator on languages, i.e., $\beta\colon 2^{X^*} \to 2^{X^*}$, $L \mapsto \beta(L)$.

Then we say $\mathcal{L}$ is closed under α if $\alpha(L_1, L_2)$ is in $\mathcal{L}$ for every L_1, L_2 in $\mathcal{L}$, and that $\mathcal{L}$ is closed under β if $\beta(L)$ is in $\mathcal{L}$ for every L in L.

The central result of this section is the following:

3 Theorem. *The family of type i languages is closed under* $\cup$, $\cdot$, *and* * *for each of* $i = 0, 1, 2, 3$.

The general strategy of the proof will be as follows: We consider one operation at a time. For definiteness, consider concatenation and the context-free languages. From G_1 and G_2 we construct a new grammar $\hat{G}$. We then observe that if G_1 and G_2 are context-free then $\hat{G}$ is also context-free, and $L(\hat{G}) = L_1 \cdot L_2$, so that $\mathcal{L}_2$, the family of context-free languages, is closed under concatenation.

PROOF. Let L_1 have grammar $G_1 = (X, V_1, S_1, P_1)$ and let L_2 have grammar $G_2 = (X, V_2, S_2, P_2)$. To simplify the later parts of the proof, we assume that the sets of variables V_1 and V_2 are disjoint. We also assume that the variable S is not in V_1 or V_2. We set $V = V_1 \cup V_2 \cup \{S\}$.

UNION: Form the grammar

$$G_3 = (X, V, S, P_3)$$

where $P_3 = \{S \to S_1, S \to S_2\} \cup P_1 \cup P_2$. Note that for $i = 0$, 1 or 2, we have that if G_1 and G_2 are of Type i then so too is G_3. In every case, clearly

$$L(G_3) = L_1 \cup L_2$$

since a derivation from S must either start with $S \Rightarrow S_1$ and then can only yield a terminal string from $L(G_1)$, or $S \Rightarrow S_2$ initiates a derivation of a string of L_2. Thus $\mathcal{L}_0$, $\mathcal{L}_1$, and $\mathcal{L}_2$ are all closed under union.

However, if G_1 and G_2 are right-linear, G_3 is *not* right-linear, because $S \to S_1$, $S \to S_2$ are not of the form $A \to bB$ or $A \to b$. To get around this difficulty we want S to initiate the first step of any G_1 derivation and *also* to initiate the first step of any G_1 derivation. We thus form

$$G_4 = (X, V, S, P_4)$$

where

$$P_4 = \{S \to w | S_1 \to w \text{ is in } P_1\}$$

$$\cup \{S \to w | S_2 \to w \text{ is in } P_2\}$$

$$\cup P_1 \cup P_2$$

Now it is clear that G_4 is right-linear when G_1 and G_2 are right-linear, and that $L(G_4) = L_1 \cup L_2$. Thus $\mathscr{L}_3$ is also closed under union.

CONCATENATION: Form the grammar

$$G_5 = (X, V, S, P_5)$$

where $P_5 = \{S \to S_1 S_2\} \cup P_1 \cup P_2$.

If G_1 and G_2 are both of Type i for $i = 0, 1, 2$, then G_5 is of the same type. If G_1 and G_2 are context-free (Type 2) then $L(G_5) = L_1 \cdot L_2$ also, because any derivation from S looks like

$$S \Rightarrow S_1 S_2 \overset{*}{\Rightarrow} w_1 w_2 \text{ where } S_1 \overset{*}{\Rightarrow} w_1 \quad \text{and} \quad S_2 \overset{*}{\Rightarrow} w_2$$

At first sight, it seems that this proof works for Type 0 and Type 1, too. However, the following example shows that "context-free" is a crucial condition to prevent G_5 from overproducing:

$$\text{Suppose } P_1 = \{S_1 \to b\}$$

$$P_2 = \{bS_2 \to bb\}$$

Then $L_1 = \{b\}$, $L_2 = \varnothing$ and $L_1 \cdot L_2 = \varnothing$. However, G_5 has the derivation

$$S \Rightarrow S_1 S_2 \Rightarrow bS_2 \Rightarrow bb$$

Thus, to make the proof work in grammars (Type 0 or 1) which have context-sensitive productions, we must stop such cross talk between the derivation from S_1 and the derivation from S_2. We do this by using distinct copies of X in these different derivations, so that no variable derived from S_1 can use part of the string derived from S_2 as context for applying a production. Let then $X' = \{x'|x \in X\}$ and $X'' = \{x''|x \in X\}$ be the two copies of X which are disjoint from each other and from X and V. Then let P_1' be obtained from P_1 by replacing each occurrence of a terminal symbol x in X by the corresponding x' in X'. Similarly, let P_2'' be obtained from P_2 by replacing each x by the corresponding x''. Then the "bleeding" of context is avoided, and we have that $L(G_6) = L_1 \cdot L_2$ when

$$G_6 = (X, V \cup X' \cup X'', S, P_6)$$

with

$$P_6 = \{S \to S_1 S_2\} \cup P_1' \cup P_2'' \cup \{x' \to x|x \in X\} \cup \{x'' \to x|x \in X\}$$

Thus Type 0 and Type 1 are each closed under concatenation.

However, there is no similar patch for Type 3 because $S \to S_1 S_2$ is not of the correct form.

In order to get around this difficulty, we modify each production in G_1 of the form $A \to b$ by placing at the right the start symbol of G_2: $A \to bS_2$. More formally we proceed as follows.

Let G_1 and G_2 be right-linear grammars. Then the grammar

$$G_7 = (X, V, S_1, P_7)$$

with

$$P_7 = \{A \rightarrow bB | A \rightarrow bB \text{ is in } P_1\} \cup \{A \rightarrow bS_2 | A \rightarrow b \text{ is in } P_1\} \cup P_2$$

is clearly right-linear, and $L(G_7) = L_1 \cdot L_2$.

ITERATION: Consider the grammar

$$S \rightarrow \lambda | S'$$

$$S' \rightarrow S_1 | S_1 S'.$$

Then S derives λ and S_1^n for $n \geq 1$. Hence the grammar

$$G_8 = (X, V_1 \cup \{S, S'\}, S, P_8)$$

with

$$P_8 = \{S \rightarrow \lambda | S', S' \rightarrow S_1 | S_1 S'\} \cup P_1$$

is context-free if G_1 is, and $L(G_8) = L_1^*$. Thus $\mathscr{L}_2$ is closed under the $*$ operation. For $\mathscr{L}_0$ and $\mathscr{L}_1$ we have the same bug as in the proof for concatenation, and to fix it we adopt the notation and analysis of the concatenation case. First we eliminate any productions of the form $S_1 \rightarrow \lambda$. Then, using the auxilliary sets of variables X', X'' to avoid cross talk, we form two copies of grammar G_1:

$$G_1' = (X, V \cup X' \cup \{S_1\}, S_1, P_1')$$

$$G_1'' = (X_1 V \cup X'' \cup \{S_2\}, S_2, P_1'')$$

Finally, we combine G_1', G_1'' with productions that build finite sequences of alternating copies S_1 and S_2 to obtain L_1^*. More specifically, let

$$G_9 = (X, V \cup X' \cup X'' \cup \{S_1, S_1', S_2, S_2'\}, S, P_9)$$

with

$$P_9 = \{S \rightarrow \lambda | S_1' | S_2', S_1' \rightarrow S_1 | S_1 S_2', S_2' \rightarrow S_2 | S_2 S_1'\} \cup P_1' \cup P_1''.$$

To construct a right-linear grammar for L_1^* from a right-linear grammar G_1 for L_1, we first introduce a new start-variable S and the rule $S \rightarrow \lambda$ to generate the empty string, we delete $S_1 \rightarrow \lambda$ if it belonged to P_1, and we add $S \rightarrow w$ for each remaining $S_1 \rightarrow w$ production to "get things started." Then for each termination production $A \rightarrow b$ of G_1 we *add* the production $A \rightarrow bS$ without deleting $A \rightarrow b$, so that, having derived a string of the form $w_1 w_2 \ldots w_j' A$ such that each of $w_1, w_2, \ldots, w_j = w_j' b$ is in L_1, we have the choices of stopping with $w_1 w_2 \ldots w_j$ or setting up $w_1 w_2 \ldots w_j S$ to initiate derivation of an even longer string of L_1^*. Thus if G_1 is right-linear, then so too is

$$G_{10} = (X, V_1 \cup \{S\}, S, P_{10})$$

with

$$P_{10} = \{S \rightarrow \lambda\} \cup (P_1 - \{S_1 \rightarrow \lambda\}) \cup \{A \rightarrow bS | A \rightarrow b \text{ is in } P_1\}$$

and $L(G_{10}) = L_1^*$.   □

To round out our discussion of closure properties of languages, we note, and make use of, certain results that will be proved in later sections.

4 Proposition. *The right-linear (Type 3) languages are closed under complement and intersection.*

PROOF. We will prove closure under complement in the next section, Proposition 2.3.10. Now, by DeMorgan's law,

$$A \cap B = \overline{\overline{A} \cup \overline{B}}$$

and thus right-linear languages are closed under intersection as well.   □

Our next result highlights some of the differences between the various language classes that make up the Chomsky hierarchy.

5 Theorem. *The context-free languages are not closed under intersection or complement.*

PROOF. Consider the languages

$$L_1 = \{a^n b^n c^m | n, m > 0\},$$
$$L_2 = \{a^n b^m c^m | n, m > 0\}.$$

Now each of L_1 and L_2 is context-free (Exercise 2). But $L_1 \cap L_2$ is not: their intersection yields the non-context-free language

$$\{a^k b^k c^k | k > 0\},$$

as shown in Example 1.3.5. For the result on the complement, apply DeMorgan's law again, as in the preceding proof.   □

It turns out that the Type 0 languages are not closed under complement; we will prove this result in Chapter 5. As for the context-sensitive languages, a recently announced result has established that this language class is indeed closed under complement.

Next, we look at the effect of substitutions on languages.

6 Definition. A map $g: X^* \to 2^{Y^*}$ is a *substitution map* if $g(\lambda) = \{\lambda\}$ and for each $n \geq 1$, $g(a_1 \ldots a_n) = g(a_1)g(a_2) \ldots g(a_n)$. That is, $g(a_1 \ldots a_n)$ is built up by concatenating the images of each of the symbols of the string under g. We define $g(L)$, where L is a language over X^* to be

$$g(L) = \bigcup_{w \in L} g(w)$$

If $g(a)$ is a singleton containing only one element of Y^* for each $a \in X$, then g is called a *homomorphism*. Intuitively, a homomorphism merely changes the names of the elements that make up strings—except that a homomorphism may also send an element $a \in X$ to λ: $g(a) = \lambda$ or to a string.

7 Example. Suppose $X = \{a, b\}$. Consider the map g with $g(a) = 0^*$, $g(b) = 1$. If $L = (aba)^*$, then $g(L) = (0^*10^*)^*$.

8 Theorem. *The context-free languages are closed under context-free substitution mappings.*

PROOF. Let $L = L(G)$ be a context-free language over alphabet X, and let g be a substitution map such that $g(x) = L_x$, a context-free language with start symbol S_x. Replace every occurrence of each x in the productions of G by the variable S_x, and add to the productions of G the productions of each grammar G_x generating an L_x, with each variable of G_x subscripted by x. The resulting grammar generates the language $g(L)$. □

9 Theorem. *The right-linear languages are closed under substitution mappings.*

PROOF. Exercise 3. □

EXERCISES FOR SECTION 2.2

1. Prove that the language $\{ww' | w, w' \in \{a, b\}^*, w \neq w', |w| = |w'|\}$ is context-free.

2. Prove that the language $\{a^n b^n c^m | n, m > 0\}$ is context-free.

3. Prove Theorem 9.

4. For $L \subset X^*$, define $\text{Init}(L) = \{z \in X^* | \text{there is a } w \in X^* \text{ and } zw \in L\}$.
   (a) What is $\text{Init}(L)$, where $L = \{a^n b^n | n > 0\}$?
   (b) Prove that if L is right-linear, then $\text{Init}(L)$ is right-linear.
   (c) Prove that if L is context-free, then $\text{Init}(L)$ is context-free.
   (d) Does $\text{Init}(L) = \text{Init}(\text{Init}(L))$? Prove this claim or give a counterexample.

5. Use closure under union to show that each of the following language is context-free:
   (a) $\{a^i b^j | i \neq j\}$
   (b) $\{a, b\}^* - \{a^i b^i | i \geq 0\}$
   (c) $\{w \in \{a, b\}^* | w = w^R\}$.

6. Suppose that L is a context-free language and R is a right-linear language. Is $L - R$ necessarily context-free? What about $R - L$? Justify your answers.

7. The right-linear languages are closed under union. Are they closed under *infinite* union? Justify your answer.

8. Prove that context-sensitive languages are closed under λ-free substitution mappings, but not under substitution mappings in general. (The substitution mapping g is λ-free if for every $a \in X$, $\lambda \notin g(a)$.)

2.3 Regular and Finite-State Languages

In this section we present an algebraic description of right-linear languages as the languages denoted by *regular expressions*. Then we introduce a machine-theoretic approach to these languages by showing that the right-linear languages are precisely those accepted by *finite-state acceptors*. The equivalence of these three characterizations is the principal result concerning right-linear languages. We prove this result, which is known as Kleene's theorem, in this section.

1 Definition. Let X be a finite alphabet. A language $L \subseteq X^*$ is *regular* if L is finite, or L can be obtained inductively using one of the following operations:

(1) L is $L_1 \cup L_2$, the *union* of L_1 and L_2, where L_1 and L_2 are regular; or
(2) L is $L_1 \cdot L_2$, the *concatenation* of L_1 and L_2, where L_1 and L_2 are regular; or
(3) L is L_1^*, the *iterate* of L_1, where L_1 is regular.

Notice that the empty set $\varnothing$, and the set $\{\lambda\}$ are both regular, being finite. The languages

(i) $\{ab\}^* \cdot \{a\}^*$
(ii) $\{a, b\}^*$

are also regular. The first consists of all strings that begin with a (possibly empty) string of alternating a's and b's, followed by a (possibly empty) block of a's. The second language consists of all finite strings of a's and b's.

From our study of closure properties in the previous section, we immediately deduce the following:

2 Proposition. *Every regular set is a right-linear language.*

3 Definition. Let X be a finite alphabet. We define *regular expressions* R and their regular set denotations $S(R)$ inductively as follows:

(i) $\varnothing$ is a regular expression with $S(\varnothing) = \varnothing$, the empty set.
(ii) λ is a regular expression with $S(\lambda) = \{\lambda\}$.
(iii) If $a \in X$, then a is a regular expression with $S(a) = \{a\}$.
(iv) If R_1 and R_2 are regular expressions then $(R_1 + R_2)$ is a regular expression with $S((R_1 + R_2)) = S(R_1) \cup S(R_2)$.
(v)   If R_1 and R_2 are regular expressions then $(R_1 \cdot R_2)$ is a regular expression with $S((R_1 \cdot R_2)) = S(R_1) \cdot S(R_2)$.
(vi) If R is regular, then $(R)^*$ is regular with $S((R)^*) = (S(R))^*$.

4 Examples

(i) $(a + b)^*$ denotes all strings over $X = \{a, b\}$

(ii) $(a^* \cdot b^*)^*$ also denotes all strings over $X = \{a, b\}$ (Why?)

(iii) $(aa + bb)^*$, which abbreviates $((a \cdot a) + (b \cdot b))^*$, stands for all finite strings of a's and b's made up by concatenating pairs of a's and pairs of b's. (We will continue to use unambiguous abbreviations.)

(iv) $aa(b^* + aaa)c + (a + c)^*$ stands for the language which is the union of three sublanguages:

    (1) all strings beginning with a doubled a, followed by any number of b's, followed by a c;

    (2) the 1 string language $\{aaaaac\}$; and

    (3) the language made up of all strings of a's and c's.

While equality between regular expressions and equality between regular sets are distinct notions, it is cumbersome to use different notations for the two concepts. So by abuse of notation we write "$=$" for both. Thus, the equation $R_1 = R_2$ does not mean that R_1 and R_2 are identical strings of symbols, but rather that $S(R_1) = S(R_2)$.

Notice that the associative laws $((\alpha + \beta) + \gamma) = (\alpha + (\beta + \gamma))$ and $((\alpha \cdot \beta) \cdot \gamma) = (\alpha \cdot (\beta \cdot \gamma))$ hold for regular expressions because these laws hold for union and concatenation: The set denoted by $((\alpha + \beta) + \gamma)$ equals the set denoted by $(\alpha + (\beta + \gamma))$. Of course, the actual regular expressions involved are different expressions. Various distributive laws hold as well. Algebraic properties of regular expressions are discussed further in the exercises.

5 Example. Consider the regular expression $(b^* + (ab)^*)$. To find a right-linear grammar for this expression, first form a grammar for b^*: $S_1 \rightarrow bS_1 | \lambda$. Then form an (ab) grammar: $S_2 \rightarrow aB_2$, $B_2 \rightarrow b$. This can be starred by adding $S_2 \rightarrow \lambda$, $B_2 \rightarrow bS_2$. Finally the full language can be generated by adding $S \rightarrow bS_1 | \lambda | aB_2$, yielding

$$S \rightarrow bS_1 | \lambda | aB_2$$

$$S_1 \rightarrow bS_1 | \lambda$$

$$S_2 \rightarrow aB_2 | \lambda$$

$$B_2 \rightarrow bS_2 | b$$

In this section we introduce a basic class of computing devices or automata called *finite-state acceptors*. (They are also discussed in Section 6.3 of *A Basis for Theoretical Computer Science*.) These devices judge the legality of strings over some finite alphabet. We shall show that the collection of languages that can be accepted by finite-state acceptors is exactly the regular languages.

6 Example. The finite state acceptor given next accepts the language $((0 + 1)0*1)*$.

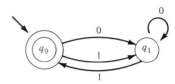

The device operates as follows on the string 1001: Processing begins in state q_0, the *start state* of the machine (indicated by the free arrow). Initially the machine scans the leftmost symbol of the input, in this case a 1. The arc from q_0 to q_1 that is labeled with a '1' indicates that when the machine is in state q_0 and is scanning a 1, it must shift into state q_1 as it advances the input scan to the next symbol. Now the machine is in state q_1 and is scanning the second symbol, a 0. In this case it reads the 0 and remains in state q_1, since the 0-arc leaving q_1 returns to q_1. Then the second 0 of the string is read, and once again the machine shifts from q_1 back to q_1. Finally, the machine scans a 1—the last symbol—and halts in state q_0. The state q_0 is an *accepting state*, as indicated by the double circle. Thus 1001 is accepted by this machine, as are 100001 and 101101. The string 1000 is rejected, since q_1 is not an accepting state.

We show that the language $((0 + 1)0*1)*$ is the acceptance set for this device, by describing all strings that send the machine from state q_0 to state q_0. Since $(0 + 1)$ represents all direct paths from q_0 to q_1, and $0*1$ describes all paths from q_1 to q_0, $(0 + 1)0*1$ represents all paths from q_0 to q_0 which visit q_0 exactly twice (once at the beginning and once at the end). Any finite number of such paths from q_0 to q_0 will produce legal strings, so the full expression is $((0 + 1)0*1)*$.

7 Definition. A *finite-state acceptor* (FSA) M with input alphabet X is specified by a quadruple (Q, δ, q_0, F) where

Q is a finite set of states
$\delta: Q \times X \to Q$ is the state transition function
$q_0 \in Q$ is the initial state
$F \subset Q$ is the set of accepting states.

Thus, the *state-graph* of an FSA has one node for each q in Q, which is so labeled; has a free arrow pointing to the q_0-node; has a double circle for a q-node just in case q is in F; and has an arc labeled x leading from the q-node to the q'-node just in case $\delta(q, x) = q'$.

The FSA of Example 6 is represented by the quadruple $(\{q_0, q_1\}, \delta, q_0, \{q_0\})$; the (finite) function δ is specified by the following chart:

δ	0	1
q_0	q_1	q_1
q_1	q_1	q_0

The behavior of the FSA, M, on a string w in X^* can be described as follows: M starts in state q_0 and scans the leftmost symbol in w. On the basis of its current state—in this case q_0—and the symbol scanned, M shifts to some new state. This transition is governed by the transition function δ given previously, which maps a state/symbol pair to a new state. The second symbol is then scanned, with M in the new state, and again a δ-transition occurs. This process continues until the entire input string has been read. If the final state is a member of F, then the string x is *accepted* by M. Given an FSA M, we shall denote by $T(M)$ the set of strings accepted by M, the set of $w \in X^*$ which sends M from q_0 to some state in F. Our goal is to characterize the set of strings which can be $T(M)$'s for some M.

8 Definition. Given an FSA M, we define by induction the function δ^*: $Q \times X^* \to Q$, such that $\delta^*(q, w)$ is the state which M will go to on input string w if started in state q_0:

Basis Step: $\delta^*(q, \lambda) = q$ for each state q in Q. If M is started in state q, then when it has received the empty input string it must still be in that same state q.

Induction Step: For each state q in Q, each input string w in X^* and each input symbol x in X, we set $\delta^*(q, wx) = \delta(\delta^*(q, w), x)$.

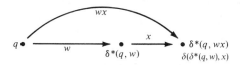

The string w sends M from state q to state $\delta^*(q, w)$, which input x then changes to state $\delta(\delta^*(q, w), x)$—but this is just the state $\delta^*(q, wx)$ to which the string wx sends M from state q.

9 Definition. The set $T(M)$ of strings accepted by the FSA $M = (Q, \delta, q_0, F)$ with input alphabet X is the subset of X^*

$$T(M) = \{w \mid \delta^*(q_0, w) \in F\}$$

comprising all those strings which send M from its initial state q_0 to an accepting state, i.e., a state in F.

We say a subset L of X^* is a *finite-state language* (FSL) if it equals $T(M)$ for some FSA M.

10 Proposition. *Every FSL is a right-linear language. Moreover, the FSL languages are closed under complement.*

PROOF. Suppose $L = T(M)$ with $M = (Q, S, q_0, F)$. Then we shall see that L is $L(G)$ for the right-linear grammar $G = (X, Q, q_0, P)$ where the set of productions is

$$P = \{q \rightarrow xq' | q' \in \delta(q, x)\} \cup \{q \rightarrow x | \delta(q, x) \in F\}$$

The reader may check that $q \overset{*}{\Rightarrow} wq'$ in G iff $\delta^*(q, w) = q'$. But then it is immediate that $q_0 \Rightarrow w$ for w in X^* just in case $\delta^*(q_0, w) \in F$, i.e., iff w is in $T(M)$.

To show that the FSLs are closed under complement, suppose that $L = T(M)$ for $M = (Q, \delta, q_0, F)$. Then $\overline{L} = X^* - L$ is $T(\overline{M})$ for the machine $\overline{M} = (Q, \delta, q_0, Q - F)$ with the complementary set of accepting states.   □

11 Example. For the FSA of Example 6 we can write the following grammar with start state q_0:

$$q_0 \rightarrow 0q_1 | 1q_1 | \lambda$$

$$q_1 \rightarrow 0q_1 | 1q_0 | 1$$

12 Theorem. *Every FSL is a regular set.*

PROOF. Let $L = T(M)$ for $M = (Q, \delta, q_0, F)$, where we suppose that $Q = \{q_0, q_1, \ldots, q_n\}$. We then let

$$R_{ij} = \{w | \delta^*(q_i, w) = q_j\}$$

be the set of all strings that carry M from state q_i to q_j. Then L is the finite union

$$L = \bigcup \{R_{0j} | j \in F\}$$

comprising all strings that carry M from q_0 to some state in F. Hence, if we can prove that each R_{ij}, with $1 \leq i, j \leq n$ is regular, we can conclude that L, being a finite union of such sets, is itself regular.

To prove the regularity of the R_{ij}'s, we introduce the auxiliary sets $R_{ij}^k = \{w | w$ leads M from q_i to q_j without passing through any of the states $q_k, q_{k+1}, \ldots, q_n$ in between$\}$. We note immediately that

$$R_{ij}^0 = \{x \in X | \delta(q_i, x) = q_j\}$$

which is finite and hence regular. We also note that

$$R_{ij}^{n+1} = R_{ij}.$$

Hence, if we can prove, by induction on k, that each R_{ij}^k is regular, we are done. The basis step is secure, and it only remains to prove the validity of the induction step.

Suppose, then, that we know $R_{i'j'}^k$ to be regular for *all* i', j' and for some $k \leq n$. We must prove that each R_{ij}^{k+1} is also regular. Now, consider (referring to Figure 1) some w in R_{ij}^{k+1}. Either the path taken when following w never reaches q_k, or else it can be broken up into segments $w_1 w_2 \ldots w_m, m \geq 2$, where w_1 takes us from q_i to q_k via $\{q_1, \ldots, q_{k-1}\}$, and w_m takes us from q_k to q_j via $\{q_1, \ldots, q_{k-1}\}$, while each w_i, for $2 \leq i < m$, takes us from q_k back to q_k via $\{q_1, \ldots, q_{k-1}\}$.

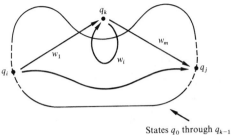

Figure 1

In the first case, we see that

$$w \in R_{ij}^k.$$

In the second case, we see that

$$w \in R_{ik}^k \cdot (R_{kk}^k)^* \cdot R_{kj}^k.$$

Thus

$$R_{ij}^{k+1} = R_{ij}^k \cup R_{ik}^k \cdot (R_{kk}^k)^* \cdot R_{kj}^k,$$

and thus, being obtained by union, dot, and star from regular sets is a regular set. Hence R_{ij}^k is regular for *all* choices of i and j.

Proceeding by induction, we then conclude that R_{ij}^{n+1} is regular, and hence that the finite-state language

$$L = \bigcup_{j \in F} R_{0j}^{n+1}$$

is also regular. $\qquad \square$

In Chapter 6 we give an alternative proof of this result by writing a set of simultaneous equations that describe the sets of strings that carry an FSA between states. Once these equations are written, they can be solved in much the same way that traditional simultaneous equations over the real numbers are solved.

13 Example. Consider M given in the following.

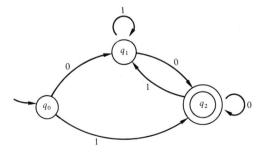

We produce $T(M)$ by the method described in the proof of Theorem 12.

Inspecting the state graph, we decide to add the state q_1 last, so we write

$$T(M) = R_{02} = A_{02} \cup A_{01} \cdot A_{11}^* \cdot A_{12}$$

where in each case A_{ij} is the set of strings which takes us from q_i to q_j without using q_1 as an intermediate state. But it is clear that

$$A_{02} = 1$$
$$A_{01} = 0 + 1 \cdot 0^* \cdot 1$$
$$A_{11} = 1 + 0 \cdot 0^* \cdot 1$$
$$A_{12} = 0 \cdot 0^*.$$

Thus we deduce that

$$T(M) = 1 + (0 + 1 \cdot 0^* \cdot 1) \cdot (1 + 0 \cdot 0^* \cdot 1)^* \cdot (0 \cdot 0^*).$$

We wish to prove that every right-linear language is a FSL by applying the proof of Proposition 10 in reverse. However, there is one immediate problem with this approach. Consider the following right-linear grammar:

$$S \to aQ_1, \qquad Q_1 \to bF \mid b$$
$$S \to aQ_2, \qquad Q_2 \to aF \mid a$$

Applying the proof of Proposition 10 in reverse leads to the following machine:

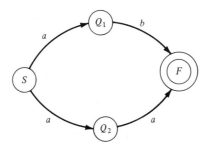

The structure obtained is not an FSA, because two arcs exiting from S have the same label, a. This violates the condition that the transition rule for an FSA be a function. When multiple transitions of this kind are present in M, we say that M is a *nondeterministic* machine. This observation motivates the next definition.

14 Definition. A *nondeterministic finite-state acceptor* (*NDA*) M with input alphabet X is specified by a quadruple (Q, δ, Q_0, F) where

Q is a finite set of states

$\delta: Q \times X \to 2^Q$ assigns to each state-input pair (q, x) a *set* $\delta(q, x) \subset Q$ of *possible* next states

$Q_0 \subset Q$ is the set of initial states

$F \subset Q$ is the set of accepting states.

We see that an NDA is just like an FSA except that there is a *set* of initial states and a *set* of possible next states. These complications mean that a string can induce multiple paths through an NDA, some ending in accepting states, others ending in nonaccepting states. We handle this ambiguity by saying $w \in X^*$ is accepted if there is *at least one* way of getting to an accepting state.

We now define $\delta^*: 2^Q \times X^* \to 2^Q$ for an NDA so that $\delta^*(p, w)$ is the set of states reachable from some state in p by applying input string w.

15 Definition. Extend $\delta: Q \times X \to 2^Q$ for an NDA to $\delta^*: 2^Q \times X^* \to 2^Q$ by:

Basis Step: $\delta^*(p, \lambda) = p$ for each $p \subset Q$.
Induction Step: $\delta^*(p, wx) = \bigcup \{\delta(q, x) | q \in \delta^*(p, w)\}$ for each w in X^*, x in X and $p \subset Q$.

We then set

$$T(M) = \{w | w \text{ in } X^* \text{ and } \delta^*(Q_0, w) \cap F \neq \varnothing\}$$

to be the set of input strings such that *at least one* path, labeled with letters of w in order, through the state graph leads M from an initial state to an accepting state.

The reader should have little trouble completing the proof of the following:

16 Proposition. *A subset of X^* is a finite-state language if and only if it is $T(M)$ for some NDA M.*

PROOF OUTLINE. (i) If $L \subset X^*$ is a finite-state language, it is $T(M_1)$ for some FSA $M_1 = (Q_1, \delta_1, q_1, F_1)$. We define an NDA $M_2 = (Q_1, \delta_2, \{q_1\}, F_1)$ in terms of M_1 by setting $\delta_2: Q_1 \times X \to 2^{Q_1}: (q, x) \mapsto \{\delta_1(q, x)\}$. Then (check the details) $T(M_1) = T(M_2)$.

(ii) If $L \subset X^*$ is $T(M)$ for the NDA $M = (Q, \delta, Q_0, F)$ we may define an FSA $M' = (Q', \delta', q_0', F')$ in terms of M by

$$Q' = 2^Q$$

$$\delta'(p, x) = \bigcup \{\delta(q, x) | q \in p\} \text{ for each } p \in Q', x \in X$$

$$q_0' = Q_0 \in Q'$$

$$F' = \{p | p \cap F \neq \varnothing\} \subset Q'.$$

Then (check the details) $T(M) = T(M')$. (See Exercise 6.)   □

17 Example. Consider the following NDA.

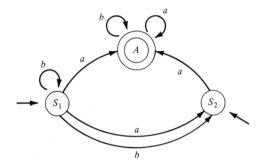

States S_1 and S_2 are start states, while state A is the only accepting state. Proposition 16 indicates that there are $2^{|Q|}$ (in this case, $2^3 = 8$) states in the equivalent FSA, but in practice it is usually not necessary to consider all of them, because many are unreachable. We take advantage of this observation by starting from the new start state and building in additional states as they are generated. Thus we begin with

	a	b
$\{S_1, S_2\}$	$\{S_2, A\}$	$\{S_1, S_2\}$

which describes the action of δ' on the initial state $\{S_1, S_2\}$ of inputs a and b, respectively.

Only one new state has been generated, so we next produce its row:

	a	b
$\{S_2, A\}$	$\{A\}$	$\{A\}$

Then we get

	a	b
$\{A\}$	$\{A\}$	$\{A\}$

and we are done. In graphical form the constructed FSA is

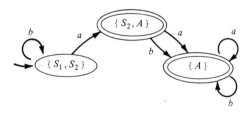

We have now developed enough machinery to prove that every right-linear language can be represented by $T(M)$ for some finite state acceptor M.

18 Theorem. *Every right-linear language L is a FSL.*

PROOF. Let $L = L(G)$ where $G = (V, X, S, P)$ is a right-linear grammar. We first construct a NDA M with input alphabet X which accepts L, i.e., $T(M) = L$. The NDA $M = (Q, S, Q_0, F)$ is defined by:

$$Q = V \cup \{q\} \text{ where } q \notin V,$$

$$Q_0 = \{S\},$$

$$F = \{q\},$$

and for all $A \in V$ and $x \in X$,

$$\delta(A, x) = \{B | B \in V \text{ and } A \to xB \text{ is in } P\} \cup \{q | A \to x \text{ is in } P\}.$$

We leave as an exercise the proof that indeed $T(M) = L$. By Proposition 16, there is a FSA M' such that $T(M') = T(M) = L$.   □

Putting together Proposition 2, Theorem 12, and Theorem 18, we conclude that the classes of right-linear languages, regular languages, and finite-state languages coincide. Our next theorem, the analogue of the pumping lemma for context-free languages, will allow us to conclude that the regular languages are properly contained in the context-free languages.

19 Theorem (The Pumping Lemma for Regular Languages). *Let M be any FSA with n states, and let $z \in T(M)$, $|z| \geq n$. Then z can be written as uvw, and the language $uv^*w \subset T(M)$. That is, for all $j \geq 0$, $uv^jw \in T(M)$.*

PROOF. Let us use the notation

to indicate that $\delta(p, a) = q$ in M. Let $z = a_1 a_2 \ldots a_k$. Then we can trace $M's$ behavior on z by writing

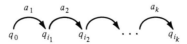

When $|z| \geq n$, at least $n + 1$ states must appear in the trace, and so some state, say r, must appear twice.

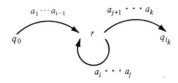

Set $z = uvw$ where $v = a_i \ldots a_j$. Then the strings uw, uvw, $uvvw$, etc., lead to the same final state of M, and so if $uvw \in T(M)$ then we have $uv^m w \in T(M)$ for all $m \geq 0$.   □

20 Example. The language $L = \{a^n b^n | n \geq 0\}$ is not regular. Suppose L were regular. Then $L = T(M)$, where M has k states, and consider $a^k b^k$. Then the theorem applies. If u is made up of all a's or all b's, then vw is $a^i b^j$ with $i \neq j$. Otherwise $u = a^p b^q$, implying that $uvvw$ has too many a-b alternations.

EXERCISES FOR SECTION 2.3

1. Describe in words the languages specified by the following regular expressions:
   (a) $(aa)^*(bb)^*$
   (b) $(a^*b^*c^*)^*$
   (c) $((a + b + c)(bb)^* + (a + b + c))^*$
   (d) $(aaa + aaaaa)^*$

2. Show that the following equivalences hold for arbitrary regular expressions R, S, and T:
   (a) $R \cdot (S + T) = R \cdot S + R \cdot T$
   (b) $(R + S) \cdot T = R \cdot S + R \cdot T$

3. Give regular expressions R and S such that
   (a) $R \cdot S = S \cdot R$
   (b) $R \cdot S \neq S \cdot R$

4. Give a right-linear grammar for the language $((a + bb)^* + c)^*$.

5. Give regular sets corresponding to the following FSAs.

   (a)

   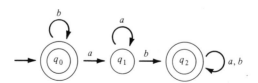

   (b)

   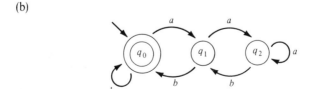

6. Complete the proof of Proposition 16. (*Hint:* Use induction on $|w|$, where $w \in T(M)$.)

7. Convert the following NDA into an FSA.

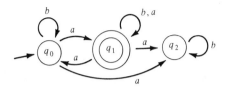

8. Assume that $R \neq \lambda$, $R \neq \emptyset$. If $\lambda \notin S(R)$ show that there is an R' such that R' does not mention λ or $\emptyset$ and $S(R) = S(R')$. If $\lambda \in S(R)$ show that $R = R' + \lambda$ where R' does not mention λ or $\emptyset$.

9. Convert the following regular sets into NDAs.
   (a) $(((11)^*0)^* + 00)^*$
   (b) $(1 + 11 + 0)^*(00)^*$

10. Use the pumping lemma for regular sets to show that the following languages are not regular.
    (a) $\{a^i b^j | i < j\}$
    (b) $\{w \in (a + b)^* | w$ has an equal number of a's and b's$\}$
    (c) the language of matched parentheses

11. Which of the following languages are regular?
    (a) $\{a^i b^j | i \neq j\}$
    (b) $\{a^i b^j | i = j^2\}$
    (c) $\{a^i b^j c^k | i = k \bmod j\}$

12. Consider left-linear grammars, that is grammars in which every production is of the form $A \rightarrow Ab$ or $A \rightarrow c$, $c \in X \cup \{\lambda\}$. A left-linear language is a language that can be generated by a left-linear grammar. Show that the left-linear languages coincide with the right-linear languages.

13. Consider languages generable by grammars of the form $A \rightarrow wB$, $w \in X^+$, $B \in (V \cup \{\lambda\})$. Show that this language class coincides with the right-linear languages.

14. (a) Let G be a right-linear grammar generating $L \subset X^*$. For $x \in X$, let $sub(x, L', L)$ stand for the replacement of every x in every string of L with the language L'. (Thus if $L = \{ab\}$, $L' = c^*$, then $sub(a, L', L) = c^* \cdot b$). Prove: L' regular implies $sub(x, L', L)$ regular.
    (b) Give an example of an L that is strictly context-free, and an L' that is regular such that $sub(x, L, L')$ is regular.

15. Repeat Example 13 with q_2 as the state added last.

CHAPTER 3

Push-Down Automata and Context-Free Grammars

3.1 Push-Down Automata

In this section we establish one of the most important equivalences in theoretical computer science: we prove that a language is context-free if and only if some push-down automaton can accept the language in a sense we shall shortly make precise. This result has practical as well as theoretical importance, because a push-down store is the basis for several algorithms used in parsing context-free languages.

A *stack* or *push-down store* is familiar to all students of computer science. A stack is a last-in first-out structure with a "push" operation which adds to the stack, and a "pop" operation which removes the top element on the stack if there is one (Figure 1).

It will turn out that the "pure" notion of a stack described informally above, augmented by the notion of *nondeterministic* state transition, provides a complete acceptor model for context-free languages. However, certain additional structure is also convenient, and the next three examples show why this is so.

1 Example. Consider the language

$$\{w \mid w \in (a + b)^*, w \text{ has an equal number of } a\text{'s and } b\text{'s}\}.$$

This language is informally acceptable by a stack using the following algorithm. Initially the stack is empty. Scan the string w from left to right, and perform the following operations, based on the current symbol at the top of the stack:

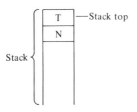

Figure 1   "Pop" pops T off stack, leaving N on top. "Push A" pushes A on top, moving T to second position.

(1) if the stack is empty and the current symbol of w is an a, put A on the stack;
(2) if the stack is empty and the current symbol of w is a b, put B on the stack;
(3) if the top stack symbol is an A and the current symbol of w is an a, push another A onto the stack;
(4) if the top stack symbol is a B and the current symbol of w is a b, push a B onto the stack;
(5) if the top stack symbol is an A and the current symbol of w is a b, pop the stack;
(6) if the top stack symbol is a B and the current symbol of w is an a, pop the stack.

You should convince yourself that this informal algorithm will correctly judge the legality of strings from this language in the following sense: a string w has an equal number of a's and b's if and only if, after processing w, the stack is empty.

Below we write a stack "program" for the informal algorithm described above. We include in our description a special symbol, Z_0, to denote the stack bottom. Then we use the notation $\langle x, D, v \rangle$ to mean "if x is the next symbol of the input string w and D is the symbol on top of the stack, then replace D by the string v." We make the convention that we write the stack as a string in such a way that the top of the stack becomes the left of the string. Then the preceding informal description may be rewritten as follows:

$$(1) \ \ \langle a, Z_0, AZ_0 \rangle$$

$$(2) \ \ \langle b, Z_0, BZ_0 \rangle$$

$$(3) \ \ \langle a, A, AA \rangle$$

$$(4) \ \ \langle b, B, BB \rangle$$

$$(5) \ \ \langle a, B, \lambda \rangle$$

$$(6) \ \ \langle b, A, \lambda \rangle$$

$$(7) \ \ \langle \lambda, Z_0, \lambda \rangle$$

Notice that this description follows the informal account given previously except that we have introduced the symbol Z_0 as stack bottom marker. We must therefore introduce the "λ-rule" (7) which, on no input, erases the bottom marker Z_0. We say a string is accepted if after the string has been completely scanned, the stack is empty.

2 Example. Consider the language $\{wcw^R | w \in (a + b)^*\}$. This language, as we shall see, is recognizable by an automaton of the kind in the last example, but a more convenient way to recognize it is to augment the stack machinery with a finite-state structure consisting of two states, one called *read* for processing the first half of the string, the other, called *match*, for the second half. The input c triggers the transition from *read* to *match*. We now use the notation $\langle q, x, D, q', v \rangle$ to mean "if the controller is in state q, x is the next symbol of the input string, and D is the symbol atop the stack, then the controller may change to state q' and replace D by the string v." We shall use the symbol Z to stand for an arbitrary stack symbol. Thus the instruction $\langle read, a, Z, read, AZ \rangle$ is shorthand for the three instructions

$$\langle read, a, Z_0, read, AZ_0 \rangle$$

$$\langle read, a, A, read, AA \rangle$$

$$\langle read, a, B, read, AB \rangle$$

Using this augmented notation we can describe our *push-down automaton* (PDA) as follows:

$$\langle read, a, Z, read, AZ \rangle$$

$$\langle read, b, Z, read, BZ \rangle$$

$$\langle read, c, Z, match, Z \rangle$$

$$\langle match, a, A, match, \lambda \rangle$$

$$\langle match, b, B, match, \lambda \rangle$$

$$\langle match, \lambda, Z_0, match, \lambda \rangle$$

Note that the PDA will halt (it has no instructions to follow) if it reads an a while B is atop the stack, etc. The following table shows the successive stack configurations on successive string segments for input $abbcbba$. When c is read, the stack contains the reverse of the string read beforehand; after reading c, the system pops the stack so long as the input symbol matches the top of the stack.

string	stack
abbcbba	Z_0
bbcbba	AZ_0
bcbba	BAZ_0
cbba	$BBAZ_0$
bba	$BBAZ_0$
ba	BAZ_0
a	AZ_0
	Z_0

This language was particularly easy for a PDA to handle because the center c informed the automaton when to "turn around" and disassemble the stack. We are not so lucky in the next example.

3 Example. Consider the language $\{ww^R | w \in (a + b)^*\}$. Here the preceding processing plan does not work because in a left-to-right scan the location of the string center is unknown. It therefore is necessary to introduce non-determinism into stack processing, essentially by "guessing" the string center whenever the center might be encountered (when two consecutive symbols are identical). We thus replace $\langle read, c, Z, match, Z \rangle$ in the preceding PDA by the λ-rule (because the "input" is just λ, not an input symbol) $\langle read, \lambda, Z, match, Z \rangle$ which says that at any time it is in state *read* the PDA may "choose" to go to state *match* without using up an input symbol or altering the stack. Here is the device:

$$\langle read, a, Z, read, AZ \rangle$$

$$\langle read, b, Z, read, BZ \rangle$$

$$\langle read, \lambda, Z, match, Z \rangle$$

$$\langle match, a, A, match, \lambda \rangle$$

$$\langle match, b, B, match, \lambda \rangle$$

$$\langle match, \lambda, Z_0, match, \lambda \rangle$$

The machine is *nondeterministic* because whenever it is in state *read* it has two possible state transitions: either reading an input symbol, *or* switching to state *match*. Clearly, a string is accepted by the PDA only if (i) the string is of the form ww^R *and* (ii) the transition to *match* is made exactly at the midpoint of reading the string.

We are thus using the same principle that we used for the nondeterministic acceptors of the previous chapter: a string is accepted if there is *at least one* choice of the possible state transitions which leads to some accepting configuration of the machine (in the present instance: empty stack after the input has been read in its entirety). In general, then, push-down automaton processing is accomplished by a set of quintuples of the form

(state, symbol, input-symbol, stack-symbol, new-state, new-stack-top)

Such quintuples express how transitions between instantaneous descriptions of the form

(state, unread input, stack)

are accomplished and in reality represent a primitive programming language. In the following we give a somewhat more formal and slightly expanded account of the model.

1 Definition. A (nondeterministic) *push-down automaton (PDA)* is a septuple $M = (Q, X, \Gamma, \delta, q_0, Z_0, F)$, where:

(1) Q is a finite set of *states*.
(2) X is a finite set of *input* symbols.
(3) Γ is a finite set of *stack* symbols.
(4) δ is the set of transitions $\langle q, x, Z, q', \sigma \rangle$ which we also write in the notation $(q', \sigma) \in \delta(q, x, Z)$ so that we may regard δ as a *transition function*:

$$\delta: Q \times (X \cup \{\lambda\}) \times \Gamma \to \text{finite subsets of } Q \times \Gamma^*$$

(5) q_0 is the *initial state*.
(6) Z_0 is the *initial stack symbol*.
(7) $F \subset Q$ is a collection of *accepting states*.

Given this machinery, we can talk about an *instantaneous description (ID)* of a PDA M. An ID for a PDA is a triple (q, w, σ) where q is a state, $w = x_1 x_2 \ldots x_n$ is a string of input symbols yet to be read with the PDA currently reading x_1, and $\sigma = Z_1 Z_2 \ldots Z_m$ is the string of symbols on the stack with Z_1 at the top and Z_m at the bottom.

Transitions map IDs to IDs nondeterministically in two ways: If M is in state q with current input symbol x and current stack element A, then

(a) If $\langle q, x, A, q', A_1 \ldots A_k \rangle$ for $x \in X$ is a quintuple of the PDA, then M may (nondeterminism!) cause the following ID transition:

$$(q, xw, A\sigma) \Rightarrow (q', w, A_1 \ldots A_k \sigma);$$

(b) If $\langle q, \lambda, A, q', A_1 \ldots A_k \rangle$ is a quintuple of the PDA, then its execution may cause the ID transition

$$(q, xw, A\sigma) \Rightarrow (q', xw, A_1 \ldots A_k \sigma).$$

That is, in (b) the machine processes the stack without advancing the input in any way. In keeping with our conventions for grammars, we use the symbol "$\overset{*}{\Rightarrow}$" in the sense *triple* $1 \overset{*}{\Rightarrow}$ *triple* 2 to mean that ID *triple* 2 can be derived after a finite sequence of zero or more transitions from *triple* 1.

5 Definition. We define the *language accepted by empty stack by a PDA M* to be

$$T(M) = \{w | (q_0, w, Z_0) \overset{*}{\Rightarrow} (q, \lambda, \lambda), \text{ for any } q \in Q\}.$$

That is, the language accepted by empty stack by a PDA is the set of all strings which, when completely processed, lead to an empty stack. Notice that F is *not* used in this definition. However, we *can* give another definition based on F of acceptance by a PDA, which we shall see is equivalent to the preceding definition.

6 Definition. Let M be a PDA. Define *the set of strings accepted by final state by M* to be

$$F(M) = \{w | (q_0, w, Z_0) \overset{*}{\Rightarrow} (q_a, \lambda, \sigma), \text{ for some } q_a \in F \text{ and } \sigma \in \Gamma^*\}$$

7 Theorem. *A language L is PDA acceptable by final state if and only if it is PDA acceptable by empty stack.*

PROOF. We show $L = T(M)$ implies that there exists a PDA M' such that $L = F(M')$. We leave the reverse implication as an exercise. Suppose $L = T(M)$. Modify M by inserting a new stack symbol Z' underneath the usual stack bottom Z_0, using the quintuple $(q_0, \lambda, Z_0, q_0, Z_0 Z')$. Add a new state q_a, and set F, the set of final states, equal to $\{q_a\}$. Now add the additional rule $(q, \lambda, Z', q_a, Z')$ to M, for each state q. The resulting machine will accept L by final state. ∎

Next we consider a modest but useful generalization of the PDA model.

8 Definition. A *generalized push-down automaton* is a PDA that behaves according to the machine description of Definition 4 except that there may be transitions in δ which read a *string* of stack symbols (instead of exactly one symbol). Thus there may be finitely many quintuples of the form

$$\langle q, a, B_1 \ldots B_k, q', C_1 \ldots C_n \rangle$$

9 Proposition. *If a language L is accepted by a generalized PDA, then L is accepted by a PDA.*

PROOF. We show how to replace every generalized transition with a finite sequence of ordinary transitions. Given a generalized quintuple of the form

$$\langle q, a, B_1 \ldots B_k, q', C_1 \ldots C_n \rangle \quad (*)$$

we generate a sequence of new intermediary transitions which guide the PDA from ID $(q, aw, B_1 \ldots B_k \sigma)$ to ID $(q', w, C_1 \ldots C_n \sigma)$. In the case of $(*)$, the following rules suffice:

$$\langle q, a, B_1, r_2, \lambda \rangle$$
$$\langle r_2, \lambda, B_2, r_3, \lambda \rangle$$
$$\vdots$$
$$\langle r_k, \lambda, B_k, q', C_1 \ldots C_n \rangle$$

The proof for quintuples of the form

$$\langle q, \lambda, B_1 \ldots B_k, q', C_1 \ldots C_n \rangle$$

is similar. ∎

In Section 3 of this chapter we will prove that the context-free languages are exactly the languages accepted by push-down automata. Half of this result, which shows that every context-free language is accepted by a push-down automaton, is based on the Greibach normal form for a context-free grammar. We establish this normal form result in Section 2. To motivate this "simulation" of context-free grammars by push-down automata, we first show how the NDA derived from a right-linear grammar can be seen as a one-state PDA.

10 Observation. Given an NDA $M = (Q, X, \delta, F)$, form the one-state PDA $\hat{M}$. We use the symbol $*$ to denote $\hat{M}$'s single state. The stack symbols of $\hat{M}$ are the states, Q, of M. Thus we can write $\hat{M}$ as follows:

$$\hat{M} = (\{*\}, X, Q, \hat{\delta}, *, q_0, \{*\})$$

the transitions $\hat{\delta}$ of $\hat{M}$ mimic δ according to the formula "treat the symbol atop the stack of $\hat{M}$ like the state of M," so that

$$(*, q') \in \hat{\delta}(*, x, q) \quad \text{iff} \quad q' \in \delta(q, x)$$

It is then clear that

$$(*, w, q_0) \overset{*}{\Rightarrow} (*, \lambda, q) \quad \text{iff} \quad q \in \delta^*(q_0, w)$$

If we then add to $\hat{\delta}$ the rules

$$(*, \lambda) \in \hat{\delta}(*, x, q) \quad \text{iff} \quad \delta(q, x) \in F$$

we deduce that

$$(*, w, q_0) \overset{*}{\Rightarrow} (*, \lambda, \lambda) \quad \text{iff} \quad \delta^*(q_0, w) \in F.$$

Thus

$$T(\hat{M}) = T(M).$$

In terms of a right-linear grammar, this says that the rule

$$A \to bB$$

yields the quintuple $\langle *, b, A, *, B \rangle$ (which we now abbreviate to $\langle b, A, B \rangle$ as in Example 1), while the rule

$$A \to b$$

yields the quintuple $\langle *, b, A, *, \lambda \rangle$. In other words, now using S in place of q_0, the derivation

$$S \overset{*}{\Rightarrow} w_1 A \Rightarrow w_1 bB \overset{*}{\Rightarrow} w_1 bw_2 = w$$

is mimicked by the PDA passing through the IDs

$$(*, w, S) \overset{*}{\Rightarrow} (*, bw_2, A) \Rightarrow (*, w_2, B) \overset{*}{\Rightarrow} (*, \lambda, \lambda)$$

Stated informally, this says that the partially processed input string contains the portion of the original string w which is unread, and the overall string is accepted iff the variable on the stack can derive that string.

The next example generalizes this to a case in which the stack contains a *sequence* of variables, rather than just a single one.

11 Example. Consider the grammar for $\{a^n b^n | n \geq 1\}$ using productions

$$S \to aSB | aB$$

$$B \to b$$

We define a one-state *nondeterministic* PDA M with the transitions

$$\langle a, S, SB \rangle$$

$$\langle a, S, B \rangle$$

$$\langle b, B, \lambda \rangle$$

Clearly, on reading the input string $a^n b^n$ we have the possible intermediate results

(1) $(*, a^n b^n, S) \overset{*}{\Rightarrow} (*, a^{n-j} b^n, SB^j)$

(2) $(*, a^n b^n, S) \overset{*}{\Rightarrow} (*, a^{n-k} b^n, B^k)$ $\qquad (k > 1)$

(3) $(*, a^n b^n, S) \overset{*}{\Rightarrow} (*, b^n, B^n)$

The derivation (1) is an intermediate stage to derive an ID of either the form (2) or (3). But (2) is a dead end—no transition is applicable—if $n > k$, while (3) is *en route* to the complete derivation

$$\langle *, a^n b^n, S \rangle \overset{*}{\Rightarrow} \langle *, \lambda, \lambda \rangle.$$

The reader can extend this analysis to see that $T(M)$ is indeed $\{a^n b^n | n \geq 1\}$.

In the preceding example, we see that we have generalized the notion of a right-linear grammar to allow productions to have an arbitrary string of variables following the initial terminal on the right-hand side.

EXERCISES FOR SECTION 3.1

1. Give a PDA which accepts the language of matched parentheses by final state.

2. Give PDAs which accept the following languages by final state:

   (a) the language $\{a^m b^m | m \leq n \leq 2m\}$;
   (b) the language consisting of all strings with twice as many a's as b's;
   (c) the language $\{w \in \{a, b\}^* | w = w^R\}$;
   (d) the language $\{a^m b^m c^n | m \geq 1 \text{ and } n \geq 1\}$.

3. Give a PDA which accepts by final state the language generated by the context-free grammar:

$$S \to aAA, A \to aS | bS | a$$

4. A λ-transition (or λ-move) in a PDA is a transition in which the input symbol is not used (case (B) in Definition 4). Show that if M_1 is a PDA which accepts language L by final state, then there is a PDA M_2 which also accepts L by final state but which makes no λ-moves.

5. Show that given a PDA M with transition function δ, one can construct an equivalent PDA M' with transition function δ' such that if $(q',\sigma) \in \delta'(q,x,Z)$ then $|\sigma| \leq 2$; i.e., M' never tries to place more than two stack symbols at a time. (M' is equivalent to M in that it accepts the same language as M, whether by final state or by empty stack.)

3.2 Normal Forms for Context-Free Grammars

In mathematics a *normal form* is a standard way of viewing some mathematical object. Thus, the fraction $\frac{1}{2}$ is a normal form for $\frac{5}{10}$, $\frac{3}{6}$, and $\frac{100}{200}$.

Normal forms play an important role in language theory, and in this section we will discuss several important normal form results for context-free grammars.

1 Lemma. *Let G be a context-free grammar. Then there exists a context-free grammar G' such that*:

(i) $L(G) = L(G')$;
(ii) *If $\lambda \notin L(G)$, then there are no productions of the form $A \to \lambda$ in G'*;
(iii) *If $\lambda \in L(G)$, then there is a single λ-production in G', $S' \to \lambda$, where S' is the start symbol of G' and S' does not appear on the right-hand side of any production in G'; and*
(iv) *There are no productions of the form $A \to B$ in G', where $A, B \in V$.*

Note that conditions (i), (ii), *and* (iii) *are exactly those of Lemma* 2.1.4.

PROOF. By Lemma 2.1.4 we may assume that the grammar G already satisfies conditions (ii) and (iii). The desired grammar G' will have the same set of variables as G, as well as the same start symbol. The new set P' of productions is obtained by first including it in all the productions of P which are *not* of the form $A \to B$ where $A, B \in V$. Then, considering all pairs A, B of variables, if $A \overset{*}{\Rightarrow} B$ according to P, we also include in P' all productions of the form $A \to w$ where $B \to w$ is a production of P and $w \notin V$.

We leave as an exercise the proof that $L(G) = L(G')$. $\square$

2 Definition. We say that a context-free grammar $G = (V, X, S, P)$ is in *Chomsky normal form* if each production is of one of the following forms:

(i) $S \to \lambda$,
(ii) $A \to BC$, where $A, B, C \in V$,
(iii) $A \to a$, where $A \in V$ and $a \in X$.

Furthermore, if $S \to \lambda$ is in P, then $B, C \in V - \{S\}$ in clause (ii).

3 Theorem. *Let G be an arbitrary context-free grammar. Then there is an equivalent grammar G', $L(G')$, in Chomsky normal form.*

PROOF. We assume that G satisfies (ii), (iii), and (iv) of the preceding lemma. For each production $A \to B_1 B_2 \ldots B_m$ in P, where $m \geq 2$, we may also assume that all the B's are variables. For suppose B_j is a terminal symbol, then we can replace it in the production with a new variable not in V, B_j', and we add the additional production $B_j' \to B_j$.

The set of productions P' of G' is obtained by first including in it all the productions in P which are of the form $A \to w$ where $|w| \leq 2$. Then, for every production $A \to B_1 B_2 \ldots B_m$ in P, where $m \geq 3$, P' will contain $(m - 1)$ new productions

$$A \to B_1 C_1$$
$$C_1 \to B_2 C_2$$
$$\vdots$$
$$C_{m-3} \to B_{m-2} C_{m-2}$$
$$C_{m-2} \to B_{m-1} B_m$$

where $C_1, \ldots, C_{m-2}$ are new variables not in V.

Note that if $m = 3$, the production $A \to B_1 B_2 B_3$ is replaced by the two productions $A \to B_1 C_1$ and $C_1 \to B_2 B_3$. It is now easy to verify that $L(G) = L(G')$, Exercise 2.   □

4 Example. Given the grammar

$$S \to ABC$$
$$C \to BaB \,|\, c$$
$$B \to b \,|\, bb$$
$$A \to a$$

we convert to Chomsky normal form by first converting terminals on right-hand sides to variables, and adding appropriate productions:

$$S \to ABC$$
$$C \to BA_1 B \,|\, c$$
$$A_1 \to a$$
$$B \to b \,|\, B_1 B_1$$
$$B_1 \to b$$

Finally we split right-hand sides with more than two variables:

$$S \rightarrow AD$$

$$D \rightarrow BC$$

$$C \rightarrow BE|c$$

$$E \rightarrow A_1 B$$

$$A_1 \rightarrow a$$

$$B \rightarrow b|B_1 B_1$$

$$B_1 \rightarrow b$$

5 Definition. We say that a context-free grammar is in *Greibach normal form* (GNF) if every production is of the form

$$A \rightarrow bW$$

where $b \in X$ while $W \in V^*$.

The key to our proof of Theorem 1 in the next section is that *every* context-free language has a grammar which is in Greibach normal form. But first we need another normal form, the *no left recursion* (NLR) normal form.

6 Lemma. *Let G be a context-free grammar. Then there is an equivalent grammar G', $L(G) = L(G')$, which has no left-recursive productions, i.e., no production of the form $A \rightarrow Av$ where $A \in V$ and $v \in (X \cup V)^*$.*

PROOF. Suppose G has only two A productions, $A \rightarrow Av|w$ where v and w are in $(X \cup V)^+$ and w does not begin with an A. Then this pair of productions derives the regular set denoted by wv^*. But we can also derive wv^* with the NLR grammar:

$$A \rightarrow wB|w$$

$$B \rightarrow vB|v$$

where B is a new nonterminal symbol. To eliminate left recursion in general, replace

$$A \rightarrow Av_1|Av_2|\ldots|Av_n|w_1|\ldots|w_m$$

(which generates the regular expression $(w_1 + \cdots + w_m)(v_1 + \cdots + v_n)^*$) with

$$A \rightarrow w_1|\ldots|w_m|w_1 B|\ldots|w_m B$$

$$B \rightarrow v_1|\ldots|v_n|v_1 B|\ldots|v_n B \qquad \qquad \square$$

One final point: notice that the new variable, B, never appears as the leftmost symbol in any production of the new grammar. This will be an important fact in the proof of the next result, which establishes that every context-free language has a Greibach normal form.

7 Theorem. *Let G be any context-free grammar. Then there is an equivalent grammar G', i.e., $L(G) = L(G')$, in Greibach normal form.*

PROOF. We fix the order of the variables in the grammar G, so that we may write $V = \{A_1, A_2, \ldots, A_n\}$. We then say that the grammar *goes uphill* if every production is of the form (1) $A_j \to A_k v$ with $j < k$; or (2) $A_j \to av$ (where a is a terminal symbol).

To prove the Griebach normal form theorem formally we proceed as follows. Assume without loss of generality that our initial grammar G has the following form: the variables of the grammar are numbered as previously, and every production is either of the form $A \to A_{i_1} \ldots A_{i_n}$ where all A_i's $\in V$, or $A \to b$, where $b \in X$. Consider the A_1 productions first. If they all go uphill—that is, they are all either of the form $A_1 \to A_j v$ with $1 < j$ or $A_1 \to av$—then we go on to the A_2 productions. Otherwise there is a left-recursive A_1 rule and we replace it using the preceding lemma. Then we go to the A_2 productions, substituting for A_1 whenever we encounter a production of the form $A_2 \to A_1 x$, and then eliminating left recursions on A_2 if any are encountered. We proceed in this fashion until we reach A_n. Then by doing back substitution top to bottom we can put G in Greibach normal form. That is, all of the A_n productions must already be in Greibach normal form and since A_{n-1} productions either map to terminals or to A_n, they may be put in Greibach normal form also, using the principle of substitution. This argument carries all the way down to A_1. Finally we apply substitution again, this time to the new variables introduced when left recursion was eliminated.   □

8 Example. Consider the context-free grammar with productions $A_1 \to A_2 A_2 | 0$ and $A_2 \to A_1 A_2 | 1$. The A_1 productions already go uphill, but the $A_2 \to A_1 A_2$ does not. Substitution produces

$$A_1 \to A_2 A_2 | 0$$

$$A_2 \to A_2 A_2 A_2 | 0 A_2 | 1$$

A_2 is now left-recursive, so we apply the no-left-recursion rule (proof of Lemma 6) to get

$$A_1 \to A_2 A_2 | 0$$

$$A_2 \to 0 A_2 | 1 | 0 A_2 Z | 1 Z$$

$$Z \to A_2 A_2 | A_2 A_2 Z$$

Now the "A" portion of the grammar goes uphill, so we apply the uphill principle (proof of Theorem 7), to get

$$A_1 \to 0 A_2 A_2 | 0 A_2 Z A_2 | 1 A_2 | 1 Z A_2 | 0$$

$$A_2 \to 0 A_2 | 0 A_2 Z | 1 | 1 Z$$

Finally we can substitute into the "Z" part of the grammar to achieve full

Greibach normal form, since no variable Z can appear in leftmost position of the right-hand side of any production. Our final grammar therefore is

$$A_1 \rightarrow 0A_2A_2 | 0A_2ZA_2 | 1A_2 | 1ZA_2 | 0$$

$$A_2 \rightarrow 0A_2 | 0A_2Z | 1 | 1Z$$

$$Z \rightarrow 0A_2A_2 | 0A_2ZA_2 | 1A_2 | 1ZA_2 | 0A_2A_2Z | 0A_2ZA_2Z | 1A_2Z | 1ZA_2Z$$

EXERCISES FOR SECTION 3.2

1. Complete the proof of Theorem 3.

2. Convert the following grammar to Chomsky normal form.

$$S \rightarrow 00A | B | 1$$

$$A \rightarrow 1AA | 2$$

$$B \rightarrow 0$$

3. Convert the grammar of Exercise 3 to Greibach normal form.

4. Convert the following grammar to Greibach normal form.

$$S \rightarrow SSS | RS | 0$$

$$R \rightarrow RR | SR | 1$$

5. Prove that any context-free grammar without λ can be transformed into an equivalent grammar in which every production is of the form $A \rightarrow x$, $A \rightarrow xB$, or $A \rightarrow xBC$, where x is an arbitrary terminal symbol and A, B, and C are arbitrary variables.

6. (Big-Little or BL Normal Form). Let G be any type 0, 1, or 2 grammar. Show that there is a grammar G' of the same type as G such that $L(G) = L(G')$ and such that every production $v \rightarrow w$ has $v \in V^*$ where w is in either V^* or X^*.

7. By first building an appropriate PDA, give a Greibach normal form grammar for the language $\{w | w$ has twice as many a's as b's$\}$.

8. Let G be a context-free grammar, $w \in L(G)$, and $|w| = n$. Give an upper bound on the length of a derivation of w in G if (a) G is in Chomsky normal form and (b) G is in Greibach normal form.

9. Show that the transformation of a context-free grammar into Chomsky normal form may square the number of productions.

10. Show that every linear language (see Exercise 2.1.10) may be generated by a context-free grammar G in which each of the productions is of the form:

$$A \rightarrow a$$

or
$$A \rightarrow aB$$

or
$$A \rightarrow Ba$$

where $A, B \in V$ and $a \in X \cup \{\lambda\}$. The grammar G is said to be in *linear normal form*.

11. Let G be a context-free grammar with no useless variables—that is, for every $A \in V$, $A \overset{*}{\Rightarrow} w$ for some $w \in X^*$. G is *self-embedding* if there exists $A \in V$ such that $A \overset{*}{\Rightarrow} vAw$, $v, w \neq \lambda$.

    a. Prove that if G is not self-embedding, then there exists a Greibach Normal Form grammar G' which is also not self-embedding such that $L(G) = L(G')$.

    b. From a, conclude that a context-free language is a regular language if and only if it can be generated by a non-self-embedded grammar.

3.3 The Equivalence Theorem

Now we turn our attention to the principal result of this chapter, the *equivalence theorem*, which states that the languages accepted by PDAs are precisely the context-free languages.

1 Theorem. *Let $L = L(G)$ for some context-free grammar G. Then L is accepted by some (in general nondeterministic) PDA. That is, $L = T(M)$ for some PDA.*

PROOF. Without loss of generality assume that G is in Greibach normal form. We shall provide an acceptor for $L(G)$ in the form of a one-state PDA M. Since M has only one state $*$ we once again abbreviate quintuples $\langle *, x, z, *, w \rangle$ to the form $\langle x, z, w \rangle$. Then associate any rule of the form

$$A \to bCDE$$

with the PDA instruction

$$\langle b, A, CDE \rangle.$$

Any rule of the form

$$A \to b$$

yields the transition

$$\langle b, A, \lambda \rangle.$$

Lambda-rules become moves with no input:

$$A \to \lambda \text{ corresponds to } \langle \lambda, A, \lambda \rangle.$$

    We must now prove that $L = T(M)$. To do this we prove the stronger result that

$$(*, w_1 w_2, S) \overset{*}{\Rightarrow} (*, w_2, \sigma)$$

(i.e., after reading w_1, M will have the string σ of symbols on its stack) holds iff

$$S \overset{*}{\Rightarrow} w_1 \sigma$$

is a valid derivation in G. Taking $w_2 = \sigma = \lambda$, we see that the preceding equivalence implies that

$$(*, w_1, S) \overset{*}{\Rightarrow} (*, \lambda, \lambda) \quad \text{iff} \quad S \overset{*}{\Rightarrow} w_1$$

which says that $w_1 \in T(M)$ iff $w_1 \in L(G)$. We now prove our stronger result by induction on the length of w_1.

Basis: For $w_1 = \lambda$ there are two cases:
(a) We have both $(*, w_2, S) \overset{*}{\Rightarrow} (*, w_2, S)$ and the corresponding $S \overset{*}{\Rightarrow} S$.
(b) If $S \to \lambda$ is a production of G, we have both $(*, w_2, S) \overset{*}{\Rightarrow} (*, w_2, \lambda)$ and the corresponding $S \overset{*}{\Rightarrow} \lambda$.

Induction: Suppose that w_1 is such that $(*, w_1 w_2, S) \overset{*}{\Rightarrow} (*, w_2, \sigma)$ iff $(S \overset{*}{\Rightarrow} w_1 \sigma)$. For convenience we write $\sigma = A\tau$, $w_2 = aw_3$. We now check our induction claim for $w_1 a$. Since

$$(*, w_1 aw_3, S) \overset{*}{\Rightarrow} (*, aw_3, A\tau) \text{ by the inductive hypothesis,}$$

we have

$$(*, w_1 aw_3, S) \overset{*}{\Rightarrow} (*, w_3, \sigma'\tau) \text{ iff } S \overset{*}{\Rightarrow} w_1 A\tau \quad \text{for some } A \text{ with } A \to a\sigma',$$
$$\text{i.e., iff } S \overset{*}{\Rightarrow} w_1 a\sigma'\tau,$$

which was to be proved.   □

2 Example. Conside the following Greibach normal form grammar for the language of matched parentheses:

$$S \to (L \mid \lambda$$

$$L \to (LL \mid)$$

The following is a leftmost derivation of the string $(()())$ in this grammar.

$$S \Rightarrow (L \Rightarrow ((LL \Rightarrow (()L \Rightarrow (()(LL \Rightarrow (()()L \Rightarrow (()())$$

We give the associated (one-state) PDA next, with designated stack-bottom symbol S.

$$\langle (, S, L \rangle$$

$$\langle \lambda, S, \lambda \rangle$$

$$\langle (, L, LL \rangle$$

$$\langle), L, \lambda \rangle$$

For one-state PDAs we can reverse the method of the previous result to find a context-free grammar that generates the language accepted by the PDA. We illustrate this first with an example.

3 Example. Consider Example 1 of Section 3.1, a PDA without state (i.e., with one state) that accepts the language of equal numbers of a's and b's. Its associated grammar (replacing Z_0 by Z) is

$$Z \rightarrow aAZ|bBZ|\lambda$$

$$A \rightarrow aAA|b$$

$$B \rightarrow bBB|a$$

We summarize this process by stating the following result:

4 Theorem. *Let M be a one-state PDA. Then there exists a context-free grammar G such that $L(G) = T(M)$.*

PROOF. To form the grammar, convert triples of the form

$$(a, B, \sigma) \text{ for } \sigma \in \Gamma^*$$

into productions of the form

$$B \rightarrow a\sigma$$

Convert triples of the form

$$\langle \lambda, B, \sigma)$$

into productions of the form

$$B \rightarrow \sigma$$

The proof that $L(G) = T(M)$ is similar to that of Theorem 1 and is left as an exercise. □

Because of Theorem 4, we need only prove that any PDA can be simulated by a one-state PDA in order to establish the full equivalence of context-free languages and the class of languages accepted by PDAs. Our next theorem establishes this result.

5 Theorem. *Let M be a PDA. Then there exists a one-state PDA M' such that $T(M) = T(M')$.*

PROOF. The idea of the proof is to design the stack machinery of M' so that M' can simulate the finite-state control of M on its stack. If p and q are states and A is a stack symbol of M, then the presence of the composite stack symbol $[pAq]$ on the stack of M' signifies "M is in state p, reading an A at the top of the stack; when A is deleted from the stack, M is in state q."

Now an instruction of the form

(α) $\qquad\qquad\qquad \langle p, a, A, q, BCD \rangle$

in M is simulated in M' by *all* triples

$$\langle a, [pA*_3], [qB*_1][*_1C*_2][*_2D*_3] \rangle$$

no matter what the choice of the symbols $*_1$, $*_2$, and $*_3$ as states of M. That is, in this case the original M quintuple is replaced in M' by all triples that bear the same stack symbol sequence and the same lead state.

A similar scheme is used to simulate the quintuples of M that read no input but nevertheless do not shorten the stack. For quintuples of the form

(β) $\langle p, a, A, q, \lambda \rangle$

that pop the stack, however, we are more restrictive, and these rules are the key to the construction. We simulate these with instructions of the form

$$\langle a, [p \, A \, q], \lambda \rangle$$

i.e., when stack shrinking instructions are applied, the stack must diminish in accordance with the result state of M as well as the start state.

We deal with the initial condition by preprocessing Z_0, the start symbol for M', with the set of instructions

$$\langle \lambda, Z_0, [q_0 \, Z_0 \, *] \rangle$$

Given an instantaneous description of M', say $(w, [pA_1p_1] \ldots [p_{k-1}A_kp_k])$, we refer to the triple $\langle p, w, A_1 \ldots A_k \rangle$ as the ID's *spine*. The spine of an M' ID is just the ID with all state information except the top state stripped off. In case $k = 0$, we define the spine of the ID (w, λ) to be $(-, w, \lambda)$.

To show $T(M) = T(M')$, we must show $T(M) \subset T(M')$ and $T(M') \subset T(M)$. We show the first containment, leaving the second as an exercise (Exercise 2). We prove that $w \in T(M)$ implies that $w \in T(M')$ by showing that if (q, w, σ) is an instantaneous description of M, then an ID of M' is produced with spine $\langle q, w, \sigma \rangle$ or, in case σ is empty, with spine $\langle -, w, \lambda \rangle$. Clearly, if w is accepted by M, it is then accepted by M'. Our proof is by induction on the number of steps required to obtain M ID (q, w, σ).

Let (q, w, σ) be an ID of M. If it is produced after one step of M, then either $\sigma = A_1 \ldots A_k$ with $k \neq 0$, in which case (q, w, σ) is a spine for all IDs of M' of the form $(w, [qA_1p_1] \ldots [p_{k-1}A_kp_k])$ for all possible combinations of states $p_1, \ldots, p_k$, or else σ is empty. In the latter case, the quintuple

$$\langle q_0, a, Z_0, q, \lambda \rangle$$

is simulated in M' by

$$\langle a, [q_0 \, Z_0 \, q], \lambda \rangle.$$

which, when applied in M' yields the correct spine.

Now assume by induction that if the M ID $(p, aw, A\sigma)$ for $A \in \Gamma$, $\sigma \in \Gamma^*$ is generable within n PDA steps by M, then $\langle p, aw, A\sigma \rangle$ is the spine of some ID of M'. Suppose an instruction of the form (α) is applicable, yielding the M ID $(q, w, BCD\sigma)$ after $n + 1$ steps. Then since by induction a spine $\langle p, aw, A\sigma \rangle$ is derivable in M', an M' ID with spine $\langle q, w, BCD\sigma \rangle$ is also derivable.

Consider now the case of a deletion rule: Suppose at step $n + 1$ (q, w, σ) is produced from $(p, aw, A\sigma)$ where $\sigma = B_1 \ldots B_k$. Then since spine $\langle p, aw, A\sigma \rangle$ stands for all combinations of M' ID's of the form

$$(aw, [p A *_1] [*_1 B_1 *_2] \ldots [*_k B_k *_{k+1}])$$

then $*_1$ can be instantiated to q and the instruction

$$(a, [p A q], \lambda)$$

produces the desired effect. This completes the proof that $T(M) \subset T(M')$. □

In Section 2.3, we proved that every nondeterministic finite-state acceptor (NDA), M, is equivalent to a deterministic finite-state acceptor (FSA), M', i.e., $T(M) = T(M')$. In Examples 1 and 2 of Section 3.1 we used deterministic PDAs, but in Example 3 of the same section our PDA used nondeterminism to "choose" when to switch from reading to matching. Was this nondeterminism necessary? More generally, is every context-free language acceptable by some *deterministic* PDA? It turns out that the answer is no: there are context-free languages that cannot be accepted by deterministic PDAs.

EXERCISES FOR SECTION 3.3

1. Complete the proof of Theorem 4.

2. Complete the proof of Theorem 5.

3. Give a one-state PDA that accepts the language of Example 3.1.2.

4. Let $L = T(M)$ for some PDA M, and suppose that as M processes L, the stack of M never grows larger than k symbols for k a fixed integer constant. Prove that L is regular.

5. A context-free grammar is *linear* if the right-hand sides of its productions contain each at most one variable symbol. A PDA is *single-turn* if whenever $(q_0, w, Z_0) \overset{*}{\Rightarrow} (q', w', \sigma') \overset{*}{\Rightarrow} (q'', w'', \sigma'') \overset{*}{\Rightarrow} (q''', w''', \sigma''')$ and if $|\sigma''| < |\sigma'|$ then $|\sigma'''| \leq |\sigma''|$; that is, once the length of the stack decreases, it never again increases. Show that a language is generated by a linear context-free grammar iff it is accepted by a single turn PDA (by final state or by empty stack).

CHAPTER 4

Parsing, Part I

In this chapter and in Chapter 7 we consider parsing, which has been the principal application of formal language theory to computer science. In compiler theory, parsing a string of terminal symbols generally means supplying a derivation tree for the string.

We begin with a discussion of the Cocke–Kasami–Younger algorithm, a procedure that will parse any context-free language in time proportional to n^3, where n is the length of the string to be parsed. Then we discuss Earley's algorithm, a procedure with running time on *deterministic languages*—that is, on languages that can be recognized by a PDA using only deterministic transition rules—in time proportional to n^2, where n is the input length. In Chapter 7 we will discuss strong $LL(k)$ grammars. These grammars can be parsed top-down with no backup by so-called recursive descent parsers. In Chapter 7 we also consider $LR(k)$ grammars. These grammars can be parsed bottom-up with no backup.

4.1 The Cocke–Kasami–Younger Algorithm

The Cocke–Kasami–Younger (CKY) algorithm is a simple procedure for recognizing strings in a context-free language. Its simplicity is achieved by requiring that the grammar be in Chomsky normal form. Because of this requirement, the derivation tree of any string will be essentially binary.

We begin with the following example.

1 Example. Consider the following Chomsky normal form grammar:

$$S \rightarrow AB|BB$$
$$A \rightarrow CC|AB|a$$
$$B \rightarrow BB|CA|b$$
$$C \rightarrow BA|AA|b$$

Now consider the top of the derivation tree for the string *aabb*. It must have one of the following forms:

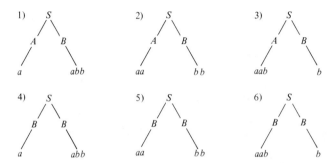

That is, the top of the tree may induce any possible split of the string into two parts which is consistent with the S productions of the grammar. For any string of length N, $N - 1$ possible divisions of the string are possible. The CKY algorithm provides an efficient way of generating substring divisions and then checking whether each substring can be legally derived. The algorithm constructs a pyramid above the string to be parsed. For the string *aabb*, the pyramid will have the following structure:

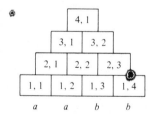

The cells of the pyramid are labeled by row and column, from bottom to top and from left to right. The cells are to be filled with variables in such a way that any entry in a cell can derive the portion of the string its pyramid "covers." More formally, a variable is to be placed in cell (i, j) just in case it is possible to derive from it i consecutive symbols of the string, starting with the jth. (Check that this coincides with the informal notion of "covering" for the preceding pyramid.) Thus, the string can be placed in $L(G)$ if the top cell of the pyramid contains the start symbol of G.

Now consider each of the six possible partial derivations given earlier. If one of them produces the string *aabb*, then the variables *A* and *B* must appear in the cells of the corresponding diagram as shown in the following:

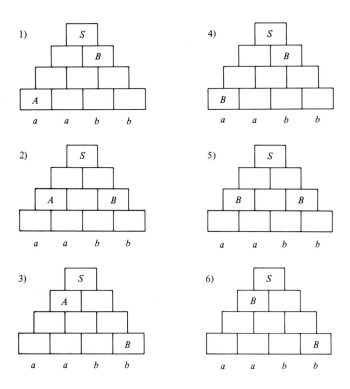

Remember again that variables in cells at height 1 can derive a string of length 1, variables in cells at height 2 can derive strings of length 2, etc. Remember also that the second subscript gives the starting position of that string. Therefore, variables in cell (4, 1) can derive the entire string starting at the first position.

The CKY algorithm fills the cells of the pyramid from the bottom up. Let's see how this works for the string *aabb*. In the following diagram we have filled in the base cells.

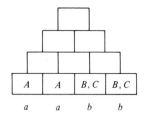

Next we consider row 2. Cells in row 2 can derive strings of length 2. For cell $(2, 1)$ we only need to look at the pairing of cells $(1, 1)$ and $(1, 2)$, so we look for productions that derive AA, which gives only C. So we place C in cell $(2, 1)$. For cell $(2, 2)$ we examine cells $(1, 2)$ and $(1, 3)$. We look for productions that produce either AB or AC, giving us S and A. For cell $(2, 3)$ we examine cells $(1, 3)$ and $(1, 4)$. Here we look for productions producing BB, BC, CB, or CC, giving us S, A, and B. With row 2 completed we have:

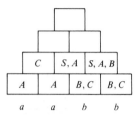

Now recall our general rule: a variable is to be placed in cell (i, j) if it can derive the substring of the given string which is of length i and starts with the jth letter of the string. Thus for $w = x_1 x_2 \ldots x_n$, A is placed in cell (i, j) for $i > j$ just in case there is a derivation $A \Rightarrow BC \overset{*}{\Rightarrow} x_j \ldots x_{j+i-1}$ for some B and C. If B derives the first m symbols and C derives the remaining $i - m$ symbols, then A is placed in cell (i, j) by pairing A with B from cell (m, j) and C in cell $(i - m, j + m)$. Note that, given i, there are $i - 1$ choices for m.

Thus for each cell in row 3 we must consider two different cell pairings. For cell $(3, 1)$ we look at cells $(1, 1)$ and $(2, 2)$ as well as $(2, 1)$ and $(1, 3)$. For the pair $(1, 1)$ and $(2, 2)$ we look for productions that produce AS or AA and find C. For $(2, 1)$ and $(1, 3)$ we look for CB or CC productions, finding A. Thus we place A and C in cell $(3, 1)$. For cell $(3, 2)$ we examine cells $(1, 2)$ and $(2, 3)$ and $(2, 2)$ and $(1, 4)$. For $(1, 2)$ and $(2, 3)$ we look for AS, AA, or AB productions, which yields S, A, and C. For the $(2, 2)$ and $(1, 4)$ pairing, we need SB, SC, AB, or AC productions, giving us S and A. So we place S, A, and C in cell $(3, 2)$ and get:

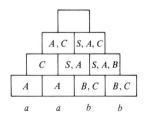

For row 4 we must consider 3 cell pairings, $(1, 1)$ and $(3, 2)$; $(2, 1)$ and $(2, 3)$; and $(3, 1)$ and $(1, 4)$. For the first pairing we look for AS, AA, or AC productions and get C. For the second pairing we look for CS, CA, or CB productions and get B. For the third pairing we look for AB, AC, CB, or CC productions and get S and A. So the string $aabb$ is in the language since S is in the cell $(4, 1)$. Using the table as a guide, we can build a parse tree for the string, which is:

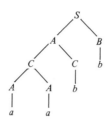

2 The CKY Algorithm. Given a context-free grammar $G = (X, V, S, P)$ in Chomsky normal form, we can decide whether $w = x_1 \ldots x_n$ is in $L(G)$ using a pyramid with base of size n as follows:

(1) Build the base cells. For $j = 1$ to n, enter an $A \in V$ in cell $(1, j)$ just in case the production $A \to x_j$ is in the grammar.
(2) Build subsequent rows: For each row $i > 1$, assume the rows $\leq i$ have already been built. Then, for $j = 1$ to $n - i + 1$, build cell (i, j) as follows: For $m = 1$ to $i - 1$, examine cell pairs (m, j) and $(i - m, j + m)$. If an entry B is in the first cell and an entry C is in the second cell, for which there is a production $A \to BC$ in the grammar, then place A in cell (i, j). The string $w \in L(G)$ iff the symbol S appears in cell $(n, 1)$.

3 Theorem. *The CKY algorithm is correct. That is, given a grammar $G = (X, V, S, P)$ in Chomsky normal form and $w = x_1 \ldots x_n$ from X^*, then A in V is in cell (i, j) of the CKY pyramid if and only if $A \overset{*}{\Rightarrow} x_j \ldots x_{j+i-1}$.*

PROOF. We use induction on i, the row number. When $i = 1$, the row in question is row 1, and the result is immediate. So suppose the result holds for all rows $< i$. If $A \Rightarrow BC \overset{*}{\Rightarrow} x_j \ldots x_{j+i-1}$, then for some $m > 0$, $B \overset{*}{\Rightarrow} x_j \ldots x_{j+m-1}$, and $C \overset{*}{\Rightarrow} x_{j+m} \ldots x_{j+i-1}$. So by the inductive hypothesis, B is in cell (m, j) and C is in cell $(i - m, j + m)$. By step 2 of the algorithm, this places A in cell (i, j). Conversely, if A is in cell (i, j), then also by step 2 there must exist variables B and C in lower cells that placed A in cell (i, j). By induction, the variables B and C must derive the two segments of w, and so we have $A \Rightarrow BC \overset{*}{\Rightarrow} x_j \ldots x_{j+i-1}$. $\square$

To aid the use of CKY as a parsing algorithm, we may replace the second sentence of (2) by the following:
    If an entry B is in cell (m, j) and an entry C is in cell $(i - m, j + m)$ and there

is a production $A \rightarrow BC$ in the grammar, then place

$$[A: (B, m, j), (C, i - m, j + m)]$$

in cell (i, j).

Clearly, by following all such pointers, starting from any S-entries in cell $(n, 1)$ we can algorithmically read off all derivation trees for w from the completed CKY pyramid.

NOTATION FOR COMPLEXITY

In computer science, it is seldom enough to establish that an algorithm is correct—one must also assess its complexity. What is its space complexity: will it run using only the available memory? What is its time-complexity: how long will it take to run?

In the present chapter, we will be concerned with time complexity. First we must decide which steps in our algorithm are "atomic." We will count each of these as consuming a single unit of time. Then we attempt to estimate how many steps are required to run the algorithm on each input. For now, our concern is parsing algorithms. We want to determine how many steps it takes to parse a string w. Rather than giving the exact time required for each specific string, we usually seek an *upper bound* $c(n)$ as a function of the length w.

4 Definition. We say that a parsing algorithm $\mathscr{A}$ has *time complexity* at most $c(n)$, where c is a function $\mathbf{N} \rightarrow \mathbf{N}$, if the parsing of each string w of length n takes at most $c(n)$ steps. In particular, $\mathscr{A}$ has *polynomial time complexity* if $c(n)$ may be chosen to be a polynomial.

Some algorithms are exponentially complex in which case we can take $c(n)$ to be of some form such as $a2^n$ for some constant a. For such algorithms there can be no polynomial $c(n)$ that bounds the algorithm's running time. Algorithms that are exponentially complex are usually considered to be too time-consuming to be practical.

5 Observations. Let $c(n) = a_k n^k + a_{k-1} n^{k-1} + \cdots + a_1 n + a_0$ be a polynomial of degree k. Consider the ratio

$$\frac{c(n)}{n^k} = a_k + \frac{a_{k-1}}{n} + \cdots + \frac{a_1}{k^{n-1}} + \frac{a_0}{k^n}$$

Clearly, as n gets large, each term after a_k gets closer to 0. Thus if we pick any constant $d > 0$, we conclude that there is a value N such that for every $n > N$ we have

$$\frac{c(n)}{n^k} < a_k + d$$

On the other hand, note that by making n sufficiently large, we can make $c(n)/n^{k+1}$ arbitrarily small.

These observations lead us to two useful notations:

6 Notation. We say that a function $c(n)$ is $O(p(n))$ (of order $p(n)$; "big oh" of $p(n)$) if there is a *fixed* constant b and an integer N such that

$$c(n) < bp(n) \quad \text{for all} \quad n > N.$$

We say that a function $c(n)$ is $o(p(n))$ (of smaller order than $p(n)$; "little oh" of $p(n)$) if for any constant b, *no matter how small*, there is an integer N, depending on b, such that

$$c(n) < bp(n) \quad \text{for all} \quad n > N.$$

Clearly if $c(n)$ is $o(p(n))$ it is certainly $O(p(n))$ but the converse is not true: $3n^3 + 2$ is $O(n^3)$ but not $o(n^3)$.

We have seen that every polynomial of order k is $O(n^k)$ and $o(n^{k+1})$. But the notation is not limited to polynomials. For example, $\sin(n)$ is $O(1)$, and $\log(n)$ is $o(n)$; while 2^n is not $O(n^k)$ for *any* k.

Note that if the time complexity of an algorithm is $O(n^3)$ we can only guarantee that doubling the size of the problem increases the time by a factor of 8; by contrast, an $O(n)$ algorithm will only double the running time for double the problem size. Thus much ingenuity has gone into seeking algorithms with a lower and lower k (not necessarily an integer) such that they solve a problem in time $O(n^k)$.

COMPLEXITY OF THE CKY ALGORITHM

In order to analyze the complexity of the CKY algorithm, we must agree on a set of primitive operations. We shall choose three:

1. Variables are paired from different cells in the matrix, forming possible right-hand sides of productions.
2. A possible right-hand side is compared against an entry in the list of actual productions.
3. A variable and a terminal are checked to see whether the variable derives the terminal.

Let us agree to assign unit cost to each of these operations. A more subtle analysis would have to consider the cost of computing indexes of cells and retrieving their entries. Notice that the number of entries in a cell is at most k, where $k = |V|$, and since G is in Chomsky normal form, there are at most k^3 productions in the grammar of the form $A \rightarrow BC$.

7 Theorem. *Let $G = (X, V, S, P)$ be a context-free grammar with k variables. Then using the CKY algorithm, $w \in L(G)$ can be decided in time proportional* *to n^3, where $n = |w|$. That is, the CKY algorithm has time complexity $O(n^3)$.*

PROOF. We establish our $O(n^3)$ upper bound by counting up the number of cells in the matrix, and then bounding the complexity of building any single cell.

First of all there are

$$n + (n - 1) + \cdots + 1 = n(n + 1)/2$$

cells in the matrix. Moreover, every cell in a row > 1 requires $a \cdot (n - 1)$ primitive operations, where a is some constant independent of n, which we arrive at as follows. A cell in a row above row 1 is formed as the union of at most $n - 1$ pairings of lower cells ($n - 1$ is achieved only by the highest, row n, cell). For each such pairing, at most k variables are paired with at most k other variables, and each of these must be checked against the at most k^3 productions in the grammar. Thus each pairing has complexity at most k^5. But k is a constant that is independent of the length, n, of the string. So if we set $a = k^5$, then the complexity of building a cell in a row above row 1 is $a \cdot (n - 1)$.

If a cell is in row 1, then we can check the terminal below it against each of the k variables. Each of these checks, by (3), requires unit cost. So row 1 cells have complexity k. Hence the overall complexity of the algorithm is bounded above by

$$k \cdot n + a \cdot (n - 1) \cdot [n(n + 1)/2] = O(n^3) \qquad \square$$

Thus the CKY algorithm is a polynomial-time algorithm for recognizing strings in a context-free language. However, the algorithm is not really practical because $O(n^3)$ is much too slow for parsing, and the constant of proportionality, a, is much too large, given that a grammar for a programming language must be converted to Chomsky normal form. Note, too, that in designing a compiler we are more interested in parsing with respect to a semantically structured grammar than in testing membership in a language in terms of a different, even though equivalent, grammar.

EXERCISES FOR SECTION 4.1

1. Set out the CKY tables to determine which of the strings *abbba, baaab, aaa, aabb* is in $L(G)$ for the grammar G of Example 1.

2. Construct the *full* CKY table with pointer for the string *baaab* for the grammar of Example 1, and use it to construct all derivation trees for the string.

3. Prove or disprove the following conjecture: Let $L(G)$ be a context-free language with the property that in every CKY chart for a string in $L(G)$, every chart entry has exactly one member. Then G is unambiguous.

4. We say that a grammar is in triple-Chomsky normal form if every production of the grammar is of the form $A \to BCD$ or $A \to a$. Modify the CKY algorithm so that it works with grammars that are written in this form. Estimate the complexity of this new algorithm.

5. (a) Modify the CKY algorithm so that it works correctly with linear context-free grammars (see Exercises 2.1.10, 3.2.10, and 3.3.5). These are a restricted class of context-free grammars, and the modified CKY algorithm should be simplified accordingly.

   (b) Show that the CKY algorithm in part (a) will parse a linear language in time proportional to n^2.

6. (a) On the basis of the CKY algorithm of this section, give an algorithm to determine the number of distinct derivations of a string w by a context-free grammar G in Chomsky normal form.

   (b) Associate a cost with each production of G. Modify the algorithm of part (a) so that it produces a minimum-cost parse of a string w. (The cost of a parse is the sum of the costs of the productions used.)

4.2 Earley's Algorithm

The CKY algorithm of the previous section was *bottom-up*, i.e., we constructed the derivation tree working from leaves *up* to the root (very unbotanical, these derivation trees). We accepted a string just if we could grow up from the string a tree with the elements of the string at its leaves and the start symbol, S, at its root. By contrast, this section is devoted to Earley's algorithm, which is *top-down*—we grow trees down from the root and seek a tree whose leaves match the given string.

Rather than visualize the parsing process in terms of trees, it will help formalize the algorithm if we think in terms of leftmost derivations. Let us consider a production $A \to v$. It is only a candidate for use in a leftmost derivation of a target string $w = x_1 \ldots x_n$ if we can find a derivation of the form

$$S \overset{*}{\Rightarrow} x_1 \ldots x_i A\delta$$

That is, A must be the leftmost variable in a sentential form in which the symbols to the left of A comprise an initial string of w. Now suppose we may decompose v as $\alpha\beta$ so that $A \to \alpha\beta$. Then consider the stage at which our leftmost derivation has expanded α but has not yet started to expand β. Again, all the terminals to the left of β must comprise an initial string of w, so that we have

(*) $$S \overset{*}{\Rightarrow} x_1 \ldots x_i A\delta \overset{*}{\Rightarrow} x_1 \ldots x_i x_{i+1} \ldots x_j \beta\delta$$

Of course, at this stage in the leftmost derivation, we don't known whether $\beta\delta$ can be expanded to complete w with the symbols $x_{j+1} \ldots x_n$. Nevertheless, our expansion of A in (*) has incrementally advanced the derivation of w from the ith symbol to the jth symbol. Such incremental derivations are a prominent feature of Earley's algorithm. The algorithm tabulates these derivations systematically and in the process builds a top-down parse of the original string. The next definition introduces the key concept in Earley's algorithm. As in the CKY algorithm, we use a matrix of cells to keep track of the parse.

1 Definition. Let $G = (X, V, S, P)$ be an arbitrary context-free grammar, and let $w = x_1 \ldots x_n$ be a string of X^*. We introduce a matrix of cells $E_{i,j}$ for $0 \le i \le j \le n$ (but $i < n$) and place the "dotted production" $A \to \alpha.\beta$ (either α or β may be empty) in $E_{i,j}$ just in case $A \to \alpha\beta$ is a production of G and there is a leftmost derivation in G of the form

$$S \overset{*}{\Rightarrow} x_1 \ldots x_i A\delta \Rightarrow x_1 \ldots x_i \alpha\beta\delta \overset{*}{\Rightarrow} x_1 \ldots x_j \beta\delta$$

Note: $A \to \alpha.\beta$ is placed in $E_{i,j}$ whether or not the preceding sentential form can be completed to yield a derivation of w. Note also that w is in $L(G)$ if and only if a dotted production of the form $S \to v.$ is placed in cell $E_{0,n}$ (why?).

2 Example. Consider the following grammar, which builds up expressions E using symbols a and b and the $+$ operator; forms equations R between such expressions; and then, from start symbol S, encloses such equations in an arbitrary number of parentheses:

$$S \to (S) | R$$

$$R \to E = E$$

$$E \to (E + E) | a | b$$

For example, the language generates $a = b$, $(a = (a + a))$, etc. To see how Early's algorithm works, we will examine a leftmost derivation of the string $a = b$ and at the same time describe the construction of the Earley matrix given in Figure 1.

i \ j	0	1	2	3
0	$S \to .(S)$ $S \to .R$ $R \to .E = E$ $E \to .(E + E)$ $E \to .a$ $E \to .b$	$R \to E. = E$ $E \to a.$	$R \to E = .E$	$S \to R.$ $R \to E = E.$
1				
2			$E \to .(E + E)$ $E \to .a$ $E \to .b$	$E \to b.$
	a	$=$	b	

Figure 1   Parse of $a = b$.

Any leftmost derivation of $a = b$ must begin with the production $S \to R$. The dotted entry $S \to .R$ in cell $E_{0,0}$ of the matrix reflects this observation. However, $E_{0,0}$ holds five other entries as well. Why?

According to Definition 1, a dotted production $A \to \alpha.\beta$ is placed in $E_{0,0}$ just in case

$$S \stackrel{*}{\Rightarrow} A\delta \Rightarrow \alpha\beta\delta \stackrel{*}{\Rightarrow} \beta\delta$$

That is, because i and j are 0, α must be empty; and the resulting dotted production $A \to .\beta$ is placed in $E_{0,0}$—again by Definition 1—just in case a sentential form with A as its leftmost symbol is derivable from S. You should convince yourself that this condition is true for all the entries in $E_{0,0}$.

Let us propose an expansion rule that we can use to generate the entries in $E_{0,0}$. First we include the S productions with a dot preceding the right-hand side. Then we expand: whenever there is a variable B just to the right of the dot for an entry in the cell, we add new dotted productions with the dot at the far left for all the B productions. After the S productions have been placed, the expansion rule accounts for the remaining entries in $E_{0,0}$.

Next look at the entries in cell $E_{0,1}$ of Figure 1. According to the conditions set out in Definition 1, both entries are valid members of this cell. The entry $E \to a.$ is valid because

$$S \stackrel{*}{\Rightarrow} E\delta \stackrel{*}{\Rightarrow} a\delta$$

is a consistent beginning for a leftmost derivation of $a = b$. Note that here, $\alpha = a$ and $\beta = \lambda$.

Similarly $R \to E. = E$ is a valid member of $E_{0,1}$ because

$$S \stackrel{*}{\Rightarrow} R\delta \stackrel{*}{\Rightarrow} E = E\delta$$

is also a valid beginning for a leftmost derivation of $a = b$.

The placement of these two entries suggest two more rules for matrix construction. The *shift rule* is as follows:

If $A \to \alpha.a\beta$ is in $E_{i,j-1}$, $a \in X$, and the symbol a matches the next terminal in the target string, then place $A \to \alpha a.\beta$ in $E_{i,j}$. The validity of this rule follows easily from the definition of dotted productions given in Definition 1. Application of the shift rule places the dotted production $E \to a.$ in $E_{0,1}$.

The shift rule tells how to place an entry in a cell with the dot shifted past a terminal symbol. To place an entry with the dot advanced past a variable, we use the *reduction rule*.

As an example of the reduction rule, consider the $E_{0,0}$ entry $R \to .E = E$, which says that $S \stackrel{*}{\Rightarrow} R\delta \stackrel{*}{\Rightarrow} E = E\delta$ is a possible leftmost derivation of the string $a = b$.

In order to place $R \to E. = E$ in $E_{0,1}$ we must establish that the passed-over E derives a valid relevant portion of the string. That is, we must establish the last derivation in the following chain:

$$S \stackrel{*}{\Rightarrow} R\delta \stackrel{*}{\Rightarrow} E = E\delta \stackrel{*}{\Rightarrow} a = E\delta$$

But the presence of $E \to a.$ in $E_{0,1}$ gives us this information. Thus we can conclude that

$$E = E\delta \overset{*}{\Rightarrow} a = E\delta$$

Hence we place $R \to E. = E$ in $E_{0,1}$.

Informally we have now seen the principal operations that make up Earley's algorithm. We now briefly describe how to complete the matrix in Figure 1.

The entry $R \to E = .E$ is placed in $E_{0,2}$ by the shift rule. Then, because the dot in this entry precedes a variable, three dotted E productions are added to $E_{2,2}$. (Why are these entries added to $E_{2,2}$ and not $E_{0,2}$?)

Next, the shift rule adds $E \to b.$ to $E_{2,3}$. Now the reduction rule applies again, this time depositing $R \to E = E.$ in $E_{0,3}$. Finally, reduction is applied once more, this time placing $S \to R.$ in $E_{0,3}$. Therefore, we can conclude that $a = b$ belongs to $L(G)$.

We now formally present Earley's algorithm. To simplify our presentation we restrict our treatment to context-free grammars without λ-productions, leaving to the reader the task of extending the algorithm to the general case (Exercise 4).

3 Earley's Algorithm. Given a context-free grammar $G = (X, V, S, P)$ without λ-productions and a string $w = x_1 \ldots x_n$ in X^*, we fill the cells $E_{i,j}$, $0 \le i \le j \le n$, $i < n$, according to the following procedure:

(1) *Initialization*: For every production of the form $S \to v$, place $S \to .v$ in $E_{0,0}$.
   Now repeat the following step until no new elements can be added to $E_{0,0}$:
(2) *Expansion*: For every variable A such that $B \to .Au$ is in $E_{0,0}$ and every production of the form $A \to v$, add $A \to .v$ to $E_{0,0}$.

Next, for each $j > 0$ in turn repeat steps (3), (4), and (5), until no dotted production can be added to $E_{0,j}, E_{1,j}, \ldots, E_{j,j}$ (column j). We construct $E_{0,j}$, $E_{1,j}, \ldots, E_{j,j}$, having already constructed $E_{i,k}$ for every i and k such that $i \le k < j$.

(3) *Shift* ("Moving the dot"): If $A \to w_1.x_jw_2$ is in $E_{i,j-1}$ then add $A \to w_1x_j.w_2$ to $E_{i,j}$.
(4) *Expansion*: If $A \to w_1.Bw_2$ is in $E_{i,j}$ for $B \in V$, add $B \to .v$ to $E_{j,j}$ for every production $B \to v$.
(5) *Reduction*: If $A \to w.$ is in $E_{i,j}$ and if $B \to w_1.Aw_2$ is in some $E_{k,i}$, then add $B \to w_1A.w_2$ to $E_{k,j}$.

The reduction step says that if $A \to w$ has successfully "covered" the symbols from i to j, and if $B \to w_1.Aw_2$ covers the symbols from k to i, then place $B \to w_1A.w_2$ in $E_{k,j}$ to record the fact that w_1A can now be expanded to cover symbols k through j.

Note that we never permit duplicate dotted productions in the same cell.

Before reading further, you should check your understanding of the algorithm by using it to rework Example 4. Label each item in Figure 1 with the stage at which you derived it, and the rule from (1) through (5) used to derive it. Check that the rules do indeed capture what we did by inspecting the leftmost derivations of initial segments.

4 Example. The table shown in Figure 2 displays the matrix for Earley's algorithm applied to the string $(a + b) = b$ using the grammar of Example 2.

5 Theorem (Correctness of Earley's Algorithm). *Given a string $w = x_1 \ldots x_n \in X^*$, Earley's algorithm places the dotted production $A \to \alpha.\beta$ in cell $E_{i,j}$ iff there is a leftmost derivation in G of the form*

(**) $$S \overset{*}{\Rightarrow} x_1 \ldots x_i A\delta \Rightarrow x_1 \ldots x_i \alpha\beta\delta \overset{*}{\Rightarrow} x_1 \ldots x_j \beta\delta$$

In particular, the string w belongs to $L(G)$ iff Earley's algorithm places a dotted production of the form $S \to v.$ in $E_{0,n}$.

PROOF. Given the motivation and presentation of the algorithm, the reader should have no trouble in confirming that every entry placed in cell $E_{i,j}$ does indeed satisfy (**) (Exercise 2).

Now we must prove that whenever the production $A \to \alpha\beta$ satisfies (**) the dotted production $A \to \alpha.\beta$ is placed in $E_{i,j}$. We proceed via three lemmas.

6 Lemma. *Suppose $B \to .v$ is in $E_{i,i}$ and suppose further that $v \overset{*}{\Rightarrow} x_{i+1} \ldots x_j$. Then $B \to v.$ is in $E_{i,j}$.*

PROOF. By induction on $j - i = k$, the length of the string derived from v. If $k = 1$, then either $B \to .x_j$, placing $B \to x_j.$ in $E_{i,j}$ by the *shift rule*; or $B \to C_1 \to C_2 \to \cdots \to C_m \to x_j$ where $C_1 = v$. In this latter case, all of

$$B \to .C_1$$
$$C_1 \to .C_2$$
$$\vdots$$
$$C_{m-1} \to .C_m$$
$$C_m \to .x_j$$

are placed in $E_{i,i}$ by repeated application of the expansion rule. Then $C_m \to x_j.$ is placed in $E_{i,j}$ by the shift rule and so by "chaining" these productions backward (repeated application of the reduction rule), $B \to C_1.$ is advanced to $E_{i,j}$.

Suppose now that the result holds for all $j - i < k$, and we consider i with $j - i = k$. Assume that v is not a singleton, for if it were, we could apply the

$i \backslash j$	0	1	2	3	4	5	6	7
0	$S \to .(S)$ $S \to .R$ $R \to .E = E$ $E \to .(E + E)$ $E \to .a$ $E \to .b$	$S \to (.S)$ $E \to (.E + E)$	$E \to (E. + E)$	$E \to (E + .E)$	$E \to (E + E.)$	$R \to E. = E$ $E \to (E + E).$	$R \to E = .E$	$S \to R.$ $R \to E = E.$
1		$S \to .(S)$ $S \to .R$ $R \to .E = E$ $E \to .(E + E)$ $E \to .a$ $E \to .b$	$R \to E. = E$ $E \to a.$					
2								
3				$E \to .(E + E)$ $E \to .a$ $E \to .b$	$E \to b.$			
4								
5								
6							$E \to .(E + E)$ $E \to .a$ $E \to .b$	$E \to b.$
7								
	(	a	+	b	)	=	b	

Figure 2   Parse of $(a + b) = b$.

chaining argument given previously. So suppose v has at least two members, say $v = z_1 \ldots z_m$ with $m > 1$, and each z_s in $(X \cup V)$. There are two cases to consider: z_1 could be x_{i+1}, or z_1 could be a variable. If $z_1 = x_{i+1}$ then $B \to x_{i+1} . z_2 \ldots z_m$ is clearly in $E_{i,i+1}$ by the shift rule. If $z_1 = C \in V$, and $C \to v' \overset{*}{\Rightarrow} x_{i+1} \ldots x_q$, then $C \to .v'$ is placed in $E_{i,i}$ by the expansion rule and $C \to v'.$ is in $E_{i,q}$ by the induction hypothesis. But then by the reduction rule $B \to C.z_2 \ldots z_m$ is in $E_{i,q}$.

We can proceed exactly in this manner for each of $z_2, z_3, \ldots, z_m$ and in this way we will place $B \to v.$ in $E_{i,j}$.   □

7 Lemma. *Suppose $A \to w_1.w_2 w_3$ belongs to $E_{i,k}$ where w_1, w_2 and w_3 are arbitrary strings in $(V \cup X)^*$, and suppose further that $w_2 \overset{*}{\Rightarrow} x_{k+1} \ldots x_j$. Then $A \to w_1 w_2.w_3$ belongs to $E_{i,j}$.*

PROOF. By induction on the length of w_2. If $|w_2| = 0$, the result follows trivially; so suppose the lemma holds whenever $|w_2| < p$, and consider the case where $|w_2| = p$, for $p > 0$. If $w_2 = x_{k+1} w_2'$ then $A \to w_1 x_{k+1} . w_2' w_3$ is in $E_{i,k+1}$ by the shift rule. Since $w_2' \overset{*}{\Rightarrow} x_{k+2} \ldots x_j$ and $|w_2'| < p$, then by the induction hypothesis, $A \to w_1 w_2.w_3$ is in $E_{i,j}$.

Suppose next that $w_2 = B w_2'$ for B a variable with $B \to u \overset{*}{\Rightarrow} x_{k+1} \ldots x_m$. Then by the expansion rule $B \to .u$ appears in $E_{k,k}$ and so $B \to u.$ appears in $E_{k,m}$ by Lemma 6. But then the reduction rule will place $A \to w_1 B.w_2' w_3$ in $E_{i,m}$ and so by the induction hypothesis of the present lemma $A \to w_1 B w_2'.w_3$ is in $E_{i,j}$.   □

8 Lemma. *Suppose $B \Rightarrow w \overset{*}{\Rightarrow} w_1 A w_2$ where w, w_1 and w_2 are arbitrary strings from $(V \cup X)^*$ and A is a variable. Suppose further that $B \to .w$ belongs to $E_{k,k}$ and $w_1 \overset{*}{\Rightarrow} x_{k+1} \ldots x_j$. Then for every production $A \to z$ we have that $A \to .z$ belongs to $E_{j,j}$.*

PROOF. By induction on the height of the derivation tree of minimal height deriving $B \Rightarrow w \overset{*}{\Rightarrow} w_1 A w_2$ (we will call such a tree a *minimal tree*). For the base case, suppose the height of this tree is 1. Then $B \to .w_1 A w_2$ is in $E_{k,k}$. By Lemma 7, $B \to w_1.A w_2$ belongs to $E_{k,j}$, and so by the expansion rule, it follows that $a \to .z$ is in $E_{j,j}$.

So, given $p > 1$, assume the result holds for all minimal trees of height $< p$. Then the minimal derivation must begin with $B \to v_1 D v_2$ where D "covers" A: i.e., $D \overset{*}{\Rightarrow} w_1' A w_2'$ and $v_1 \overset{*}{\Rightarrow} x_{k+1} \ldots x_m$ for some strings v_1, v_2, w_1', and w_2' in $(V \cup X)^*$. Here, $w = v_1 D v_2$. Then by the basis step all productions $D \to .u$ belong to $E_{m,m}$, and since the minimal derivation $D \overset{*}{\Rightarrow} w_1' A w_2'$ now has height $< p$, the result follows by the induction hypothesis.   □

COMPLETION OF PROOF OF THEOREM 5. The following hypotheses restate condition (**) where w_1 and w_2 are strings in $(V \cup X)^*$ and $A \to w_1 w_2$ is a production of the grammar:

$$(1) \quad S \Rightarrow v \overset{*}{\Rightarrow} w_3 A \delta$$

$$(2) \quad w_3 \overset{*}{\Rightarrow} a_1 \ldots a_i$$

$$(3) \quad w_1 \overset{*}{\Rightarrow} a_{i+1} \ldots a_j$$

We wish to conclude that these conditions place $A \to w_1.w_2$ in $E_{i,j}$. By initialization, $S \to .v$ is in $E_{0,0}$ and so Lemma 8 applies, placing $A \to .w_1 w_2$ in $E_{i,i}$. By (3) and Lemma 7, $A \to w_1.w_2$ belongs to $E_{i,j}$. $\square$

The analysis of the complexity of Earley's algorithm is beyond the scope of this volume. Instead we refer the reader to A. V. Aho and J. D. Ullman, *The Theory of Parsing, Translation, and Compiling. Volume 1: Parsing* (Prentice-Hall, Inc., 1972), Section 4.2.2, "The Parsing Method of Earley," where the following results are either proved or established through a series of exercises:

9 Theorem. (i) *If the underlying grammar is unambiguous, then Earley's algorithm can be executed in $O(n^2)$ reasonably defined elementary operations when the input is of length n.*

(ii) *In all cases, Earley's algorithm can be executed in $O(n^3)$ reasonably defined elementary operations when the input is of length n.*

EXERCISES FOR SECTION 4.2

1. Use Earley's algorithm to show that the string $(a = (a + b))$ is generated by the grammar of Example 4, but that $((a = (a + b)$ is not.

2. Show that every entry placed in cell $E_{i,j}$ by Earley's algorithm satisfies condition (**).

3. Use Earley's algorithm to parse the string **"if** p **then if** q **then** a **else** b**"** using the ambiguous grammar presented at the beginning of Section 1.4.

4. Extend Earley's algorithm to arbitrary context-free grammars, possibly containing λ-productions.

CHAPTER 5

Turing Machines and Language Theory

In this chapter we relate the theory of formal languages to the abstract theory of computing. How can we characterize the set of functions computed by computer programs? This question—which functions are realizable by algorithms, and which are not?—has its most direct roots in the work of Alan Turing in the 1930s. Using what is now called the Turing machine model, Turing showed that certain natural problems in computing cannot be computed by any algorithm, real or imagined. Actually, Turing showed only that these problems are not calculable specifically by Turing machines; later investigations by other researchers led to the generally held belief that Turing computability is synonymous with computability in any other sufficiently powerful algorithm specification system.

In our companion volume on computability theory, *A Programming Approach to Computability*, we develop an algorithmic specification system based on a PASCAL fragment called the language of **while**-programs. In that volume we examine the concept of computability in detail and show, among other things, that Turing computability and **while**-program computability coincide. In this volume we have chosen to base our work on the original Turing machine model because Turing machines are much closer in conception to the symbol processing methods present in language theory.

We begin this chapter by introducing Turing machines and then presenting Turing's principal results, the universality of Turing machines and the socalled unsolvability of the halting problem. In its most general form, the latter result says that no algorithm can correctly decide whether an arbitrary computer program halts on an arbitrary input. We then show how to encode this result into language theory. Using this encoding we prove a number of results, including the assertion that no algorithm can correctly decide whether an arbitrary context-free grammar is ambiguous.

5.1 Turing Machines

A Turing machine T may be pictured (Figure 1) as a finite-state control equipped with an external storage device in the form of a finite tape that can be extended indefinitely in both directions.

As shown in Figure 1, the tape is divided into squares. Each square of the tape may be blank or may hold any one symbol from a specified finite tape alphabet X. For convenience, a special symbol (here, the letter B) not in X is reserved to denote a blank square on the tape. Thus, in Figure 1, all squares are blank except for five, which hold the symbols a or b. The finite-state control is coupled to the tape through a read/write head. At any given instant, the head will be scanning one square of the tape and the finite-state control will be in one state. Depending on this state and the symbol in the square scanned by the tape head, the machine will, in one step, do the following:

(1) Enter a new state of the finite-state control;
(2) Overwrite a symbol on the scanned square (it is possible to overwrite with the same symbol and so leave the tape square unchanged, or overwrite with a B and "erase" the tape symbol);
(3) Either shift the head left one square or right one square, or not shift it at all.

Since the tape alphabet is finite and the finite-state control has finitely many states, the operation of a Turing machine can be completely described by a finite set of quintuples of the form (old state, symbol scanned, new state, symbol written, direction of motion). In the following example we use mnemonic names for states.

1 Example. The following Turing machine accepts the set $\{a^{2n} | n > 0\}$ of even-length strings of a's: when started in state "even" on the leftmost a, with initial tape configuration $Baaa \ldots aaaB$ it will terminate in state "accept" just in case the number of a's is even. The directional indicators R and N specify "move right one square" and "no move," respectively. (An L specifies a left move.) At any instant this machine is in state "even" if it has seen an even-length string of a's. Here is the machine:

$$(\text{even } a \text{ odd } B \ R)$$

$$(\text{odd } a \text{ even } B \ R)$$

$$(\text{even } B \text{ accept } B \ N)$$

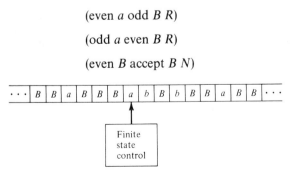

Figure 1   A Turing machine: The tape can be extended indefinitely by adding B's (blanks) at either end.

When it first reaches a B it switches to the "accept" state if it has seen an even number (including 0) of a's but it will otherwise halt because there is no quintuple to guide its next move.

By contrast, the machine with quintuples

$$(\text{even } a \text{ odd } B \; R)$$

$$(\text{odd } a \text{ even } B \; R)$$

$$(\text{even } B \text{ accept } B \; N)$$

$$(\text{odd } B \text{ odd } B \; R)$$

has the property that on even-length strings of a's it halts in the accept state, but on odd-length strings it runs forever. This illustrates how Turing machines are really different from FSAs. In particular, Turing machine computations can give rise to undefined results—computations that never terminate. Such behavior is not possible with FSAs.

As with FSAs, we can give an equivalent representation of the quintuples in Example 1 using a state table, as follows:

| old state | symbol scanned | |
	a	B
even	odd B R	accept B N
odd	even B R	

In order to make the behavior of a machine deterministic (a condition we shall shortly relax), we require that whenever two quintuples have the same (old state, state symbol scanned) combination, then the two quintuples are identical.

We can now give the formal definition of a Turing machine and the language it accepts.

2 Definition. A Turing machine M is a quintuple $M = (Q, X, q_0, q_a, \delta)$, where

Q is a finite set of states;
X is a finite tape alphabet. We write $X' = X \cup \{B\}$ for X augmented by a distinguished tape symbol B, the blank symbol;
q_0 is the distinguished start state;
q_a is the distinguished accept state;
δ is the state transition (partial) function,

$$\delta \colon Q \times X' \to Q \times X' \times \{L, N, R\}$$

subject to the condition that $\delta(q_a, x)$ is not defined for any x in X'. As we have seen, we may represent δ by a list of quintuples, with $(q \, x \, q' \, x' \, D)$ in the list just in case $\delta(q, x) = (q', x', D)$.

In Figure 1, we can describe the configuration of the Turing machine M by specifying

(i) the string w_1 printed on the tape to the left of the read/write head;
(ii) the current state q; and
(iii) the string w_2 on the tape, starting at the read/write head and moving right.

We may summarize this in the string

$$w_1 q w_2$$

which is called an *instantaneous description* (ID) of M. Note that w_1 and w_2 are *not* unique—the string $w_1 w_2$ must contain all the nonblank squares of M's tape, but it can be extended by adding blanks at either end: $1q_0 1B11$ and $BB1q_0 1B11BBB$ are two ways of writing the same ID.

Turing machines have tapes, read/write heads, etc., and cannot obviously be described using the machinery of language theory. Instantaneous descriptions, on the other hand, are really strings of symbols and as such can be manipulated by a formal grammar. We link language theory and the Turing machine notion of computability later in the chapter by showing how a grammar rewriting an instantaneous description can faithfully track the transitions of a Turing machine computation.

3 Definition. Let M be a Turing machine with state set Q and augmented tape alphabet X'. Then an instantaneous description (ID) of M is a string $w_1 q w_2$ from the set $(X')^* \cdot Q \cdot (X')^*$. Each ID $w_1 q w_2$ is equivalent to the ID's $Bw_1 q w_2$ and $w_1 q w_2 B$.

We say M *halts on* $w_1 q w_2$ if neither $w_1 q w_2$ nor $w_1 q w_2 B$ is of the form $w_1 q x w$ for a pair (q, x) for which $\delta(q, x)$ is defined. If M does not halt for $w_1 q w_2$, we define how IDs are transformed by the action of M. We first write $w_1 q w_2$ as $\bar{w}_1 y q x \bar{w}_2$ where $\bar{w}_1 y = w_1$ if $w_1 \neq \lambda$, and $\bar{w}_1 = \lambda$, $y = B$ otherwise; and $x \bar{w}_2 = w_2$ if $w_2 \neq \lambda$, and $x = B$, $\bar{w}_2 = \lambda$ otherwise. We then define the *next-ID relation*

$$w_1 q w_2 \Rightarrow w_1' q' w_2'$$

as follows:

(i) if $\delta(q, x) = (q', x', L)$ then

$$w_1' q' w_2' = \bar{w}_1 q' y x' \bar{w}_2$$

(ii) if $\delta(q, x) = (q', x', N)$ then

$$w_1' q' w_2' = \bar{w}_1 y q' x' \bar{w}_2$$

(iii) if $\delta(q, x) = (q', x', R)$ then

$$w_1' q' w_2' = \bar{w}_1 y x' q' \bar{w}_2.$$

The reader should check that this definition does indeed capture our intuition about the meaning of L, N, and R.

As usual, we extend $\Rightarrow$ to form its reflexive, transitive closure $\overset{*}{\Rightarrow}$ so that

$$\alpha \overset{*}{\Rightarrow} \alpha'$$

means that M can transform ID α to ID α' through 0 or more successor ID transitions as described in (i)–(iii) previously.

For a Turing machine M over alphabet X, we define $T(M)$, the acceptance set of M, to be the set of strings w over X such that if w is written on a blank tape and M begins processing in state q_0 on the leftmost symbol of w, then M eventually halts in the accept state. Formally.

4 Definition. The *acceptance set* $T(M)$ of a Turing machine M over the alphabet X is the set

$$T(M) = \{w \in X^* | \text{there exist } w_1, w_2 \in (X')^* \text{ such that } q_0 w \overset{*}{\Rightarrow} w_1 q_a w_2\}.$$

5 Example. The following Turing machine accepts the language $\{a^n b^n c^n | n > 0\}$. Once again we have used mnemonic state names in our machine description.

(i) When the machine reads an a, it replaces it by $\#$, then moves right to find a b:

> (start a del-b $\#$ R)
>
> (del-b a del-b a R)
>
> (del-b $\#$ del-b $\#$ R)

Note that it will halt if it encounters a c while in state del-b.

(ii) When the machine, in state del-b, reads a b, it replaces it by $\#$, then moves right to find a c:

> (del-b b del-c $\#$ R)
>
> (del-c b del-c b R)
>
> (del-c $\#$ del-c $\#$ R)

(iii) If it finds a c (not having been halted by an a en route), it then replaces it by a $\#$ and moves left to find an a, whereupon it repeats the whole cycle from (i).

> (del-c c seek-a $\#$ L)
>
> (seek-a $\#$ seek-a $\#$ L)
>
> (seek-a b seek-a b L)
>
> (seek-a a del-b $\#$ R)

(iv) However, it no a remains, if must then check that all b's and c's have been deleted. Only in this case will it accept the initial string:

$$(\text{seek-a } B \text{ check } B \ R)$$

$$(\text{check } \# \text{ check } \# \ R)$$

$$(\text{check } B \text{ accept } B \ N)$$

Notice that if a string is not of the form $a^n b^n c^n$, then the accept state is never reached, and in fact the machine will come to rest in some other state. For example, on input b^{10}, the machine never leaves the start state.

EXERCISES FOR SECTION 5.1

1. Write a Turing machine that accepts the language consisting of all strings of a's and b's in which there are more a's than b's.

2. Develop subroutines for adding and deleting a square from a Turing machine tape. More specifically.
   (a) Design a Turing machine program with a start state q_0 and halting state q_a, and with "end-marker" tape symbols \$ and ¢ such that, for all strings w_1, w_2 of tape-symbols not containing \$ or ¢ we have

   $$\$w_1 q_0 w_2 ¢ \overset{*}{\Rightarrow} \$w_1 q_a B w_2 ¢$$

   (b) Design a Turing machine program with a start state q_0' and halting state q_a', and with "end-marker" tape symbols \$ and ¢ such that, for all strings w_1 and $x w_2$ of tape-symbols not containing \$ or ¢, we have

   $$\$w_1 q_0' x w_2 ¢ \overset{*}{\Rightarrow} \$w_1 q_a' w_2 ¢$$

   (*Hint*: In each case, the machine must operate by cycling to move one symbol at a time, working between its initial position and one of the end-markers.)

3. Carefully define a PDA with two stacks. Then prove the following result: Any Turing machine can be simulated by some PDA with two stacks.

5.2 The Universal Turing Machine

So far we have only thought of Turing machines as language acceptors. It is also important to use these machines as devices that compute numerical functions, that is, map $\mathbf{N}^k \to \mathbf{N}$. Let us agree, then, to code the set of natural numbers in unary notation. Thus the code for 0 is 1, the code for 1 is 11, 2 is 111, 3 is 1111, etc. We write $\hat{n}$ to stand for n coded in unary.

Using unary notation, we can give a number-theoretic semantics for Turing machines. That is, given a Turing machine over an alphabet that includes the symbol 1, we can say how to interpret the behavior of a Turing machine so that it can be thought of as a device that computes a number-theoretic function. As we shall see, a single Turing machine according to our conventions actually computes a (different) number-theoretic function $\mathbf{N}^k \to \mathbf{N}$ for every arity k.

$$\cdots B \; q_0 \quad \Big| \; 1 \; 1 \; \cdots \; 1 \; \Big| \; B \; \Big| \; 1 \; \cdots \; 1 \; \Big| \; B \; \cdots \; B \; \Big| \; 1 \; \cdots \; 1 \; \Big| \; B \; \cdots$$

$$\leftarrow n_1 + 1 \rightarrow \qquad \leftarrow n_2 + 1 \rightarrow \qquad \leftarrow n_k + 1 \rightarrow$$

Figure 2   Initial instantaneous description for computing $\varphi_M^{(k)}(n_1,\ldots,n_k)$.

1 Definition. A Turing machine M computes a function $\varphi_M^{(k)}$ of arity k as follows. On input $(n_1,\ldots n_k)$, n_1, $\ldots n_k$ are placed on M's tape in unary, separated by single blanks (Figure 2). M's head is placed over the leftmost 1 of $\hat{n}_1$, and M's finite state control is placed in q_0. In other words, M has initial ID

$$q_0 \hat{n}_1 B \hat{n}_2 B \ldots B \hat{n}_k$$

If and when M finishes processing, the 1's on the tape are counted up (they may not appear in a block) and their total is the value of $\varphi_M^{(k)}(n_1,\ldots,n_k)$. If M never halts, we say $\varphi_M^{(k)}(n_1,\ldots,n_k)$ is *undefined*. We refer to $\varphi_M^{(k)}$ as the (k-ary) *semantics* of M.

2 Example. The Turing machine with no quintuples (i.e., it halts whatever the ID) computes the successor function $\varphi^{(1)}(n) = s(n) = n + 1$. However, the reader should check that, as a calculator for a two-variable function, it computes the function $\varphi^{(2)}(x, y) = x + y + 2$.

Our next example reveals a good deal more about the nature of Turing machine computations.

3 Example. Consider the following Turing machine.

$$(q_0 \; 1 \; q_1 \; 1 \; R)$$

$$(q_1 \; 1 \; q_0 \; 1 \; R)$$

$$(q_1 \; B \; q_1 \; B \; R)$$

This Turing machine computes the following one-variable function

$$\varphi_M^{(1)}(n) = \begin{cases} n + 1, & \text{if } n \text{ is odd;} \\ \bot, & \text{otherwise} \end{cases}$$

where the notation "$\bot$" is an abbreviation for "is undefined." That is to say, here M's semantics is a *partial function*: on some inputs, a value is returned, but on others—in this case, all even arguments—the function is undefined. This phenomenon is an undeniable fact of life in computer science. There are perfectly legal programs in any (sufficiently rich) programming language which fail to return values on some or possibly all inputs. That is why the corresponding mathematical notion of a partial function is the appropriate setting for the abstract theory of algorithms and the functions they compute.

4 Definition. A *partial function* is a function that may or may not be defined on all its arguments. Specifically, whereas an "ordinary" function $g: A \rightarrow B$ assigns a value $g(x)$ in B to each x in A, a partial function $\varphi: A \rightarrow B$ assigns a value $\varphi(x)$ only to x's in some subset $\text{dom}(\varphi)$ of A, called the *domain of definition* of φ. If $x \notin \text{dom}(\varphi)$, we say that φ is undefined or unspecified at that value. Note that we shall refer to A as the domain of $\varphi: A \rightarrow B$, and B as the codomain. The set $\{\varphi(x) | x \in \text{dom}(\varphi)\}$ is called the *range* of φ and is denoted $\varphi(A)$. If $\text{dom}(\varphi) = A$, i.e., φ assigns a value in B to every $x \in A$, then φ is called a *total function*.

The least defined partial function is the empty function. It is written $\perp$, with $\text{dom}(\perp) = \varnothing$, so that $\perp(n) = \perp$ for all n in $\mathbf{N}$. The most defined partial functions are the total functions, such as the successor function, $s(n) = n + 1$, for all n in $\mathbf{N}$. In between there are partially defined partial functions, such as the function computed in Example 3.

5 Definition. A partial function $\varphi: N \rightarrow N$ is *Turing-computable* if it is $\varphi_M^{(1)}$ for some Turing machine.

Now in an obvious way we can talk about the PASCAL-computable number-theoretic functions, the FORTRAN-computable number-theoretic functions, etc. What is the relationship between these classes? It turns out that these function classes coincide with the Turing-computable functions. After half a century of detailed studies of various computing systems and the functions they compute, it has been found that the computable functions are invariant across a wide range of different defining mechanisms—each formal system studied to date has been shown to compute either all the Turing-computable functions or some subset of them. This led the mathematical logician Alonzo Church to formulate *Church's thesis*, which holds that *all* sufficiently powerful computing mechanisms define the same class of computable functions. But Church's thesis is a hypothesis and not a theorem, because we can't prove theorems about computing systems that differ from any we have yet thought of! (For a more detailed discussion of this and other points of computability theory, the interested reader should see *A Programming Approach to Computability* by Kfoury, Moll, and Arbib.)

A familiar concept in computer science is that once a computer, say M, is sufficiently "general-purpose," a program written for any other machine may be recoded to yield a program for M which will compute the same function. Here we present a 1936 result due to A. M. Turing that antedates the digital computer by almost a decade and yet conveys the essential idea of the previous sentence: namely, that there exists a Turing machine U which is *universal*, in the sense that the behavior of any other machine M may be encoded as a string $e(M)$ such that U will process any string of the form $(e(M), w)$ just as w would be processed by M; diagramatically, we mean

$$\text{if} \quad w \overset{*}{\underset{M}{\Rightarrow}} w' \quad \text{then} \quad (e(M), w) \overset{*}{\underset{U}{\Rightarrow}} w'$$

Before constructing our universal machine U, we need to show how to enumerate all Turing machine descriptions. What will this enumeration be like? Because a Turing machine is presented as a list of quintuples, our enumeration will be a listing of lists of quintuples. Moreover, the enumeration will be done in an *effective* way, so that given an integer k we may algorithmically find our k^{th} Turing machine, and given a Turing machine, we may algorithmically find k, its position in the enumeration.

We begin by describing a special version of the "empty program," a Turing machine which is undefined for all arities and all inputs.

$$(q_0 \ 1 \ q_0 \ 1 \ R)$$

$$(q_0 \ B \ q_0 \ B \ R)$$

Clearly, on any input involving only 1's and blanks, this program runs forever. We denote this machine as M_R, since it always moves right.

Next, we give our systematic listing of all Turing machines:

$$M_0, M_1, \dots, M_k, \dots$$

assuming that the tape symbols are just B and 1, and all states are coded in the form q_k where k is a natural number written in decimal notation. Here machine M_k is determined as follows. Let us associate an ASCII-like code with every symbol that can appear in a quintuple. The following chart gives one such matchup.

100000	0
100001	1
100010	2
$\vdots$	
101001	9
110000	q
110001	1 (the tape symbol)
110010	B
111000	R
111001	L
111010	N

Using this code we can associate a binary string with any quintuple simply by concatenating the 6-bit strings associated with each symbol. For example:

$$(q_2 \ 1 \ q_{11} \ B \ L)$$

$$110000 : 100010 : 110001 : 110000 : 100001 : 100001 : 110010 : 111001$$

$$q \qquad 2 \qquad 1 \qquad q \qquad 1 \qquad 1 \qquad B \qquad L$$

(The colons ":" are not part of the string—they are just inserted to aid human scanning.)

Once we have established a coding for quintuples, we can code a full Turing machine unambiguously by concatenating the bit strings of its individual quintuples. The resulting concatenated string, interpreted as a binary number, is the code of the Turing machine. (Exercise: Why is it that we can decode this code unambiguously to recover the set of quintuples?) The natural number n is a *legal machine description* if it corresponds to a set of quintuples which are deterministic in the sense described in the previous section.

Thus we have the following listing of Turing machines,

6 $$M_0, M_1, \ldots, M_n, \ldots$$

where M_n is the machine with binary code n, if n is a legal machine description, and our fixed machine M_R for the empty function otherwise. We call n the *index* of machine M_n. (Notice that M_R appears very frequently in the list.)

Besides our listing of the Turing machines we can talk about a listing of the Turing-computable functions. The one-variable Turing-computable functions are enumerated as follows:

7 $$\varphi_0, \varphi_1, \ldots, \varphi_n, \ldots$$

where φ_n is the one-variable function $\varphi_{M_n}^{(1)}$ computed by Turing machine M_n.

Before presenting and proving our principal result, we make several observations about our Turing machine listing **6** and computable-function listing **7**. First of all, our listing demonstrates that there are only countably many Turing machines and associated functions. Second, notice that every Turing-computable function appears infinitely often in our φ_n listing. This is so because, given any machine M_n, consider M_n's state set, Q. For any state p not in Q (and there are infinitely many of these), consider the machine consisting of M_n and the additional quintuple $(p \ B \ p \ B \ N)$. This new machine, say M', does exactly what M_n does since state p can never be reached. But for each choice of state p, M' has a different index.

Finally, notice that we have only listed the one-variable functions computed by our Turing machines. We can give a listing as well for the k-variable functions for any $k > 1$. We write these as $\varphi_n^{(k)}$.

We now turn to the universal Turing machine U. First, we will show that there is no loss of generality in simulating machines having only two symbols, say B and 1, in their input alphabet. Second, we will outline a universal machine $\hat{U}$ which uses three tape heads to accomplish its task. Finally we will show that $\hat{U}$ may be replaced by a single-headed U, albeit at a great increase in computation time.

Our first stage is a simple one. If we replace each symbol of a large tape alphabet by a distinct string of m symbols from $\{1, B\}^m$, we may simulate our original machine M by a new machine $\hat{M}$ which simulates a single move of M in at most $3m$ moves. Think of the machine's tape as divided into blocks of m consecutive squares, so that each block contains the code for a single symbol of the original alphabet X' (and with B^m coding the B of X'). Then the new machine $\hat{M}$ starts its cycle of simulation sitting on the leftmost square of a block. It then reads right m steps to infer the encoded X' symbol. Next it stores the desired new state of M and completes its cycle by moving its head not at all (to simulate an R); m squares left (to simulate an N); or $2m$ squares left (to simulate an L). Since it is straightforward to encode this process in quintuple form (Exercise 7) we have the following:

8 Lemma. *Given any Turing machine* $M = (Q, X, q_0, q_a, \delta)$ *and any one-to-one map* $h: X \to \{1, B\}^m$ *which sends B to B^m, there exists a Turing machine*

$$\hat{M} = (\hat{Q}, \{1\}, q_0', q_a', \hat{\delta})$$

which simulates M in a precise sense as follows: there exists an encoding $h_q: Q \to \hat{Q}$ *of states of M such that for every ID $\alpha q \beta$ of M we may obtain the encoding of its successor in $3m$ steps by the use of $\hat{M}$; that is*

$$\delta_M(\alpha q \beta) = \alpha' q' \beta' \Rightarrow \delta_{\hat{M}}^{3m}(h(\alpha)h_q(q)h(\beta)) = h(\alpha')h_q(q')h(\beta')$$

PROOF. For uniformity we make $\hat{M}$ "cycle" for m steps when simulating an N move, and for $2m$ steps when simulating a right move so that simulation of one M-step always takes $3m$ $\hat{M}$-steps.   □

We now show how any Turing machine M with a 2-symbol alphabet can be simulated by a Turing machine $\hat{U}$ which has three heads with which it can scan the tapes (which we can conveniently represent as three tracks of a single tape), as shown in Figure 3.

The idea is that H_1 sits on the first track of the tape in just the way that the head of M sits on its binary tape. $\hat{U}$ then consults the quintuples of M, using H_2 to read their encoding on track 2, to command H_1's behavior on track 1; it uses H_3 on track 3 for subsidiary computations required for

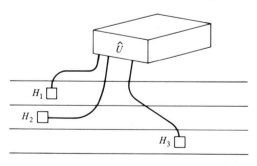

Figure 3

repositioning H_2 at the correct quintuple encoding for the next cycle of the simulation.

Although the details can be filled in in many ways, they are not very illuminating, and so will be omitted. The reader may find a complete and elegant characterization of a Universal Turing machine in Chapters 6 and 7 of Marvin Minsky's book *Computation: Finite and Infinite Machines*, Prentice-Hall, 1967. Here let us content ourselves with indicating how a multihead machine can be reduced to a single-head machine.

Suppose that we have a machine with p heads scanning a single tape on which are printed symbols from the alphabet X'. At any time the control box will be in a state q of Q and will receive as input the p symbols scanned by its heads, i.e., an element of $(X')^p$. The output of the control unit is an element of $((X')^p \times M) \cup \{halt\}$, where $M = \{I | I$ is a possible instruction to the heads to move at most distance $1\}$. Thus, the Turing machine is specified as usual by quintuples

$$q_i x_j q_l x_k I$$

with the sole difference that the x's and I's are "vectors," and that we must employ a convention to resolve conflicts if two heads try to print different symbols on a single "square." A computation of such a machine starts with the assignment of a state to the control unit and the assignment of initial positions for the heads. As usual, it is assumed that the tape has at most finitely many nonblank squares. Then computation proceeds in the usual fashion, stopping when and only when no quintuple beginning with $q_i x_j$ is applicable.

Our task is to show that any such computation can be simulated, in a suitable sense, on an "ordinary" Turing machine (i.e., with $p = 1$) such that the number of steps required for such a simulation is bounded.

9 Lemma. *Suppose a generalized Turing machine $\hat{M}$ has* (a) *p scanners on a single 1-dimensional tape or* (b) *one scanner on each of p tapes. Then $\hat{M}$ may be simulated by an ordinary Turing machine M in such a way that a single step of $\hat{M}$, when the active portion of its tape is n squares long, may be simulated by M at most $2n + 2p$ steps.*

PROOF. Let our p-head machine $\hat{M}$ have alphabet X' on its single tape in case (a). In case (b) we simply lay the p tapes side by side to form a single tape with alphabet $X' = X'_1 \times \cdots \times X'_p$ where X'_j is the alphabet on tape j. In either case, we simulate $\hat{M}$ with a 1-head machine M with alphabet $(X' \times S_p \cup \{*\})$ where S_p is the set of all subsets of $\{1, 2, \ldots, p\}$, and * marks the left-hand end of the tape. We use the enlarged tape alphabet to encode the placement of the heads on the tape. For example, if head 1 and head 3 of $\hat{M}$ scan symbol x_2, we encode this fact on M's tape by using the compound symbol $(x_2, \{1, 3\})$ instead of x_2. We write x for $(x, \varnothing)$.

More formally, if $\hat{M}$ has the nonblank tape

$$x_1 x_2 \ldots x_m$$

with head j scanning x_{i_j}, we let M have its head scan the leftmost square of

$$(x_1, A_1)(x_2, A_2)\ldots(x_m, A_M)$$

where

$$A_k = \{\,j \mid k = i_j\,\}$$

Now suppose that $\hat{M}$'s tape alphabet consists of the digits 1 through 9, and suppose that we abbreviate the symbol $(x, \varnothing)$ of M by writing just x. Then we encode the $\hat{M}$ configuration

3   1, 2

↓   ↓

3 7 4 1 2 9 1 3 4 2 2

with the M configuration

↓
* 3 7 4 (1,{3}) 2 9 (1,{1,2}) 3 4 2 2

Thus, the head information is encoded on the tape. It is now necessary to design the logic so that this information is updated after each cycle of the simulation. The control box of M can be thought of as containing p registers which are to hold the p scanned symbols of $\hat{M}$.

In a simulation, M moves ↓ right from * until it has filled all p registers. It then "knows" the new settings required and moves ↓ left, changing squares appropriately, until it has returned to * (which may have to be displaced one square left).

If the rightmost square in which the second symbol is non-null is n squares to the right of *, then the simulation will take at most $2n + 2p$ steps, the extra $2p$ steps being required to simulate the worst possible repositioning of heads.   □

Note that this result supports Church's thesis that any algorithm coded in any computer language, real or imagined, is in principle programmable by Turing machines. For, imagine that a Turing machine tape is a word of memory—in fact, an unbounded word of memory. Now suppose we consider a 512K tape/head machine. If we imagine that tape 1 is used as an infinite mass storage, and the other tapes are merely regarded as words of memory, then we will have a machine with a bizarre but serviceable machine language and an infinite memory. Such a language is adequate for realizing an algorithm written in any computer language.

We have thus established, in outline, the following result:

10 Theorem (The Universal Machine Theorem). *There exists a universal Turing machine U such that*

$$\varphi_U^{(2)}(x, y) = \varphi_x^{(1)}(y)$$

The *universal machine theorem* states that there is a single Turing machine which, when regarded as a device that calculates a two-variable function, acts as an interpreter: it treats its first argument as a program code and applies that code to its second argument. Note that Turing published this result in 1936 (when he was 24 years old), about 8 years before the first programmed electronic computer was built, and long before an actual interpreter was ever designed.

EXERCISES FOR SECTION 5.2

1. Write a Turing machine that computes $\max(n, m)$.

2. Write a Turing machine that computes $f(x) = 2 * x$.

3. Given two partial functions $\psi: N \to N$ and $\mu: N \to N$, we say that $\psi \leq \mu$ if $\mu(n)$ is always defined when $\psi(n)$ is, and if, when they are both defined, they have the same value. We write $\psi < \mu$ if $\psi \leq \mu$ but $\psi \neq \mu$.
   (a) Give functions ψ and μ such that $\psi < \varphi < \mu$, where φ is the $\varphi_M^{(1)}$ of Example 3.
   (b) Give an infinite descending chain of partial function: that is, give a sequence of partial functions

$$\varphi_1 > \varphi_2 > \cdots > \varphi_k > \cdots .$$

4. Put the proof of Lemma 8 in formal quintuple form for the case $m = 2$.

5.3 Nondeterministic Turing Machines and Linear Bounded Automata

With our experience in reducing multi-head Turing machines to single-head machines, we can build up to the result that nondeterministic Turing machines accept the same sets as deterministic Turing machines. First, let us say that a string w is accepted by a 2-tape, 2-head Turing machine if, when w is written on tape 1 and head 1 scans the leftmost symbol of w in state q_0, and tape 2 is initially blank, M eventually halts in its accepting state q_a. It then follows from Lemma 5.2.9 that

1 Theorem. *Let L be a language that can be recognized by a k-tape, k-head Turing machine. Then $L = T(M)$ for some 1-tape, 1-head Turing machine.*

2 Definition. *A machine M is called a* nondeterministic Turing machine *if it includes quintuples that specify multiple moves for a given state/symbol pair, i.e., if it is defined as in Definition 5.1.2 except that now δ maps $Q \times X'$ to subsets of $Q \times X' \times \{L, N, R\}$.*

Clearly, we can define IDs, transitions, and halting IDs as before. We say that a nondeterministic Turing machine M *accepts* a string w if there exists *some* chain of transitions of the machine on input w that reaches a halting ID which includes the accept state q_a. This definition of acceptance is in keeping with the definition of nondeterministic acceptance we gave for nondeterministic finite-state acceptors. Our next result shows that, as with FSAs, nondeterminism adds no new computational power to the original deterministic model. (This is *not* true of all machines we study—nondeterministic push-down automata (Section 3.3) *are* more powerful than deterministic push-down automata.)

3 Theorem. *Let M be a nondeterministic Turing machine that accepts L. Then there exists a deterministic machine M' that also accepts L.*

PROOF. We construct a 2-tape, 2-head deterministic machine M' which acts as an interpreter for M. On its second tape M' keeps a record of the possible active IDs of M at any instant of its computation. For example, if M includes the q_0 instructions

$$(q_0 \, a \, q_i \, b \, R)$$

$$(q_0 \, a \, q_j \, c \, L)$$

then after one instruction execution on input aa, tape 2 of M' looks like

$$\# \, b \, q_i \, a \, \# \, q_j \, B \, c \, a \, \#$$

Here we have used the symbol $\#$ to separate ID encodings. Suppose that at some instant tape 2 looks like

$(*)$ $\qquad\qquad\qquad \# \, ID_1 \, \# \, ID_2 \, \# \cdots \# \, ID_n \, \#$

Then M' operates to form the next set of IDs by processing $(*)$ left to right. When processing the jth block, M' either

(1) deletes the block altogether if no transitions are possible; or
(2) replaces the block by a new finite set of ID blocks that represent the possible new transitions of M on the given ID.   □

4 Theorem. *A language is type 0 if and only if it is the acceptance set of some Turing machine.*

SKETCH OF PROOF.
   (i) Suppose $L = T(M)$ for some Turing machine M. We exploit the extremely local character of Turing machine transitions (one major reason why Turing machines are so unpleasant to program) to write grammar rules that rewrite one instantaneous description of M to its successor description.
   We begin by writing a (context-sensitive) grammar for the set of strings

$$w \, \$ \, q_0 \, w \, \mathyen$$

where w is any string over the tape alphabet of M (see Exercise 3). $\$$ is a new, special terminal symbol, and $\not{c}$ is a special variable of the grammar. The string to the right of the $\$$ represents the initial ID of the computation on w.

Now that we have represented the initial ID as a symbol string, our next job is to devise grammar rules that match machine transitions. Suppose there is a quintuple of the machine of the form

$$(q_0 \ a \ q_j \ c \ R)$$

Then include in the grammar the production

$$q_0 \ a \rightarrow c \ q_j$$

so that the grammar's replacement rules faithfully track the machine's transitions for moves to the right. We leave left move and no move transitions to the reader. We also need rules of the form

$$\$ \ q_i \rightarrow \$ \ B \ q_i$$

$$q_i \ \not{c} \rightarrow q_i \ B \ \not{c}$$

for each $q_i \in Q$ to allow the ID to "grow" as the active portion of the tape grows.

Of course if the machine halts, we must clean up the tape so that we are left only with the accepted string. So suppose string w is accepted by the Turing machine, and state q_a (that is, variable q_a) is derived as part of some sentential form. Notice that q_a must be located to the right of the $\$$ symbol. Therefore, we add rules to the grammar that allow q_a to "gobble up" all other symbols to the right of $\$$. Specifically, for any terminal symbol a, add rules

$$q_a \ a \rightarrow q_a \quad \text{and} \quad a \ q_a \rightarrow q_a$$

and also add

$$\$ \ q_a \ \not{c} \rightarrow \lambda$$

Thus, if w is such that M's computation allows it to halt in state q_a, the preceding productions will allow us to rewrite $w \$ w_1 q_a w_2 \not{c}$ as the string w, placing w in $L(G)$. However, if M never reaches state q_a, the symbol $\not{c}$ must remain on the right end, and therefore the final string is not in $L(G)$.

(ii) Conversely, suppose that $L = L(G)$, where G is a Type 0 grammar. Here the equivalence of deterministic and nondeterministic Turing machines is of great value. To build a machine M that accepts L, design the transitions of M so that it nondeterministically attempts to parse the string inscribed on the tape. That is, start M in q_0 at the left end of the string. Move the head left-to-right across the string. If it is ever discovered that some sequence of symbols corresponds to the right-hand side of a production, M will nondeterministically insert the left-hand side of the rule at the corresponding point in the tape. Rules that continue the processing as if the rule just cited never applied must also be present. Further details are left to the reader. $\square$

We now consider *linear bounded automata*. These devices are nondeterministic Turing machines with the restriction that the read/write head is not permitted to leave the tape squares that hold the original input. We shall show that these machines are precisely the acceptors for context-sensitive languages.

5 Definition. A *linear bounded automaton* (*LBA*) is a nondeterministic Turing machine which must operate within the space defined by the original placement of the input on the tape.

6 The Alphabet Lemma. *Let L be a language such that for any $w \in X^*$, membership of w in L can be decided by Turing machine M using $k \cdot |w|$ tape squares, where k is a positive integer independent of w. Then L can be recognized by some other Turing machine M' using only $|w|$ tape squares.*

PROOF. This important result is just the converse of Lemma 5.2.8. It is proved by extending the base alphabet so that a block of several "old" symbols constitute a "new" symbol, in effect redefining the meaning of a tape square. More specifically, given base alphabet X, form the alphabet $(X')^k$ made up of sequences of k symbols from X or B. Then given string x on the tape, first concentrate it into $|w|/k$ tape squares by coding the contents of each successive block of k consecutive tape squares into a single tape square. Then mimic the behavior of the M on this new symbol set. The resulting machine will use only $|w|$ tape squares.   □

Thus, the *alphabet lemma* states that if some Turing machine can recognize a language L in space that is a linear function of the length of an input string, then some other Turing machine can in fact recognize L within space exactly equal to the length of the input. The linear relationship identified in the lemma accounts for the name *linear bounded automata*.

We assume, when convenient, that the Turing machine input is bounded on the left by a $ symbol, and on the right by a ¢. We assume that these symbols are nonerasable and that the head may not move past them.

7 Theorem. *If $L = T(M)$ for some LBA M, then L is generable by some context-sensitive grammar.*

PROOF. Let us assume that $\lambda \notin L$. For if $\lambda \in L$, then we use the theorem for $L - \{\lambda\}$ and carefully add a $S \rightarrow \lambda$ production after the fact. To establish the result assuming $\lambda \notin L$, we follow the proof of the analogous theorem for unrestricted Turing machines, Theorem 4. First we write a context-sensitive grammar for the language

$$\{w \$ q_0 w \, ¢ \,| w \in X^*\}$$

The string to the right of the $ symbol is of length $|w| + 2$. Since the machine in this case is an LBA, the IDs will never get longer than this. However, at the termination of the grammar's simulation of M, we cannot erase all symbols to the right of \$, as we did in the general case of Theorem 4, because the erasure rules are not context-sensitive. Instead, if q_a appears, we do the following:

(1) First, convert all symbols to the right of \$ to a new symbol, say "/", so that we are left with the following string:

$$w \ \$ \ /^n \ q_a \ \cent$$

(2) If $w = x_1 x_2 \ldots x_n$, use context-sensitive rules to build the following string:

$$r / x_1 / x_2 / \ldots / x_n \ \# \ q$$

(where r and q are new terminal symbols);

(3) Erase all occurrences of r, $/$, $\#$, and q.

We leave the programming details of the first two of these tasks to the reader. The third step preserves context-sensitivity because of Theorem 9 below.   ☐

Our next result builds on our study of homomorphisms in Chapter 2 to refine our understanding of context-sensitive languages. Often in writing context-sensitive grammars the most natural approach is to use a system of special marker symbols to guide the grammatical processing. And, as in the preceding proof, a grammar is written that is almost right, but leaves unnecessary marker symbols in strings of the language, markers that apparently cannot be removed except by an obvious but illegal lambda-rule. The next theorem states that limited erasing rules that eliminate such markers are indeed legal, so long as markers remain "sparse" in a language's strings.

8 Definition. A homomorphism $h: X \to Y$ is *k-limited* on $L \subset X^*$ if for all w in L, $h(w)$ erases (that is, sends to λ) at most k consecutive symbols of w. Thus the map $h(a) = a$, $h(b) = \lambda$ is 2-limited on the language $(ab)^*$ but is not k-limited for any k on the language $\{a^n b^n \mid n > 0\}$.

While the context-sensitive languages are not closed under arbitrary homomorphisms, they are closed under k-limited homomorphisms.

9 Theorem. *The context-sensitive languages are closed under k-limited homomorphisms.*

SKETCH OF PROOF. Let h be a homomorphism $h: X \to Y$ which is k-limited on $L = L(G)$, where $G = (X, V, S, P)$ is context-sensitive grammar. Suppose G is in *BL* normal form (see Exercise 3.2.6), and suppose $S \overset{*}{\Rightarrow} W \overset{*}{\Rightarrow} w$, where W is the final string of pure nonterminals in the derivation of w.

Suppose W looks like $AABBACCCCCC \ldots ABCDE \ldots ABBA$. The idea of the proof is to segment W into blocks of size i, $r \leq i < 2r$, where r is greater than k, and r is also greater than the length of any right-hand side (and therefore greater than the length of any left-hand side) of any production in

G. For example, with $r = 4$ the string *W* given previously might be segmented as

10 $[AABB][ACCCCCC]\ldots[ABCDE]\ldots[ABBA]$

Now suppose for simplicity that each capitalized variable is replaced in $L(G)$ by the corresponding lowercase letter, and if *a* is an arbitrary terminal, then $h(a)$ is either *a* itself or λ. Then if we treat a bracketed block of variables as a single variable in the grammar generating $h(L)$, and if we send $[X]$ to $h(x)$ where *X* is a string of variables and *x* is the corresponding string of terminals, then $h(x)$ is guaranteed to be nonempty since bracketed expressions are more than *k* symbols long.

Of course, we must also guarantee that the $h(L)$ grammar with bracketed variables as its nonterminals correctly tracks the successive sentential form transitions of *G*. Suppose for example that in the original grammar there is a rule $BA \rightarrow CCC$. This should induce the new rule $[ABBA] \rightarrow [ABCCC]$, which will account for one transition possible from the last block of our example string **10**. But the $BA \rightarrow CCC$ rule has other effects. Because *BA* straddles the first two blocks, the following rule must also be present in the grammar.

$$[AABB][ACCCCCC] \rightarrow [AABC][CCCC][CCCC]$$

Our rule of thumb, then, when a new bracketed expression must be introduced is as follows: the bracketing will split off *r*-sized chunks until this is no longer possible, leaving a chunk of size $i, r \le i < 2r$ on the right end. Notice also that the case where a rule of *G* falls across a bracketing boundary is the situation that places a context-sensitive production in the grammar for $h(L)$.

Now we list more precisely the rules for the $h(L)$ grammar $G' = (Y, V', S', P')$, where $G = (X, V, S, P)$ and *h* is a homomorphism $h: X \rightarrow Y$. The following grammar is complete and takes into account the special-case initial transitions when block sizes may be small.

$V' = \{[W] \mid W \in V^+, |W| < 2r\}$, where *r* is greater than the maximum of *k* and the length of the longest right-hand side of any production in *G*.
   $S' = [S]$
   $P' =$

(1) if $S \rightarrow \lambda$ is in *P*, or if there is some string $w \in L(G)$ such that $h(w) = \lambda$, then place $[S] \rightarrow \lambda$ in the grammar G' for $h(L)$.
(2) if $W \Rightarrow W'$ in *G* and $[W], [W'] \in V'$, place $[W] \rightarrow [W']$ in P'.
(3) if $W \Rightarrow W'$ in *G* and $[W] \in V', [W'] \notin V'$, place $[W] \rightarrow [W_1][W_2]$ in P' where $|W_1| = r, r \le |W_2| < 2r$ (because W' is not in V' but is produced in one step from *W*) and $W_1 W_2 = W'$.
(4) if $W_1 W_2 \Rightarrow Z_1 \ldots Z_j$, $r \le |W_1|, |W_2| < 2r$, and $|Z_i| = r$ for $i < j$, and $r \le |Z_j| < 2r$, for $W_1, W_2, Z_1, \ldots, Z_j \in V$, then place $[W_1][W_2] \rightarrow [Z_1] \ldots [Z_j]$ in P'.

(5) $[W] \rightarrow u$, if u is the result of replacing each variable of G in W with its corresponding terminal symbol, and then applying the homomorphism h to this string of terminals.

The preceding grammar is clearly context-sensitive. We leave it as an exercise to show that it generates the language $h(L)$. □

11 Example. The following grammar generates all strings of the form $\{wmwr \mid w \in \{0, 1\}^+\}$. The symbols m and r should be regarded as terminal marker symbols which identify the center and right ends of the string. The homomorphism h that erases them and leaves all other terminals fixed is clearly k-limited for $k = 1$. Hence the language $\{ww \mid w \in \{0, 1\}^*\}$ is context-sensitive. Here is the grammar.

$$S \Rightarrow Amr$$

$$Am \Rightarrow 0mT_0 \mid 1mT_1$$

$$T_0 0 \Rightarrow 0T_0$$

$$T_0 1 \Rightarrow 1T_0$$

$$T_1 1 \Rightarrow 1T_1$$

$$T_1 0 \Rightarrow 0T_1$$

$$T_0 r \Rightarrow 0r \mid 0Dr$$

$$T_1 r \Rightarrow 1r \mid 1Dr$$

$$1D \Rightarrow D1$$

$$0D \Rightarrow D0$$

$$mD \Rightarrow Am$$

EXERCISES FOR SECTION 5.3

1. Describe a Turing machine that, on an input of n a's leaves the nth prime number of a's on an auxiliary tape (e.g., when 0 a's are input, 2 a's are output; when 1 a is input, 3 a's are output)

2. Describe a branch-on-zero instruction from your favorite assembly language, and show how a multitape Turing machine could be made to simulate this instruction.

3. Show that the language

$$\{w\$q_0 w\dcent \mid w \in \{1, B\}^*, \$, q_0, \dcent \text{ are fixed symbols}\}$$

is context-sensitive.

4. Suppose $L \subset X^*$ and $a \notin X$. Suppose further that the language $L/a = \{a\ w_1\ a\ w_2\ a \ldots a\ w_n\ a \mid w_1 w_2 \ldots w_n \in L\}$ is context-sensitive. Prove that L is context-sensitive.

5.4 The Halting Problem and Undecidability

Having demonstrated the power and flexibility of effective computation, as captured in Turing machines, we now show that even this power and flexibility have limits. A very practical question for the computer scientist is, Will this program process data in the way intended? Even more fundamental, perhaps, than determining whether or not a program is correct is telling whether or not it will terminate at all, let alone with the right answer. Rephrasing the last question in terms of Turing machines, then, we have the following:

The Halting Problem. Given a Turing machine M_n with semantics φ_n and a data string w, is $\varphi_n(w)$ defined? That is, does M_n ever halt if started in state q_0 scanning the leftmost square of w?

It would be pleasant if we could find some universal halting tester which, when given a number n and a string w, could effectively tell us whether or not the machine M_n will ever halt if its input data is w. However, Theorem 3 will tell us that no such machine exists. Before we give the formal proof that *no* effective procedure can solve the halting problem, let us see why the *obvious* procedure fails. Let us simply take the universal machine U and run it on the tape $(e(M_n), w)$. When it halts, control is transferred to a subroutine that prints out a 1 to signify that the computation of M_n on w does indeed halt. But when can control be transferred to a subroutine that prints a 0 to signify that M_n never halts with data w? After a billion years of simulation by U, are we to conclude that M_n will *never* halt on w, or that, with the passage of another eon or two, computation will halt? This approach clearly fails. But to prove that *all* approaches that attempt to decide termination algorithmically will necessarily fail is a far more subtle task, and to prove it we must use a descendant of the diagonal argument used by Cantor (cf. Section 5.3 of *A Basis for Theoretical Computer Science*) to show that no enumeration could include all the real numbers.

To foreshadow our study of the halting problem, we first provide a result that relates our enumeration of Turing machines to our discussion of numerical functions. Recall that a Turing-computable function ψ is total if it returns an output for every input.

2 Theorem. *There is no total Turing-computable function g which enumerates the total Turing-computable functions in the following sense: The Turing-computable function ψ is total if and only if ψ equals the function $\varphi_{g(n)}$ computed by $M_{g(n)}$ for some n.*

PROOF. In this proof we shall make use of the notion that, if we can outline an effective procedure for computing a function, then in fact it is Turing-computable. (Recall our discussion of Church's thesis in Section 5.2.)

If g is a total Turing-computable function, then the following is an effective procedure for computing a *total* numerical function which we will call h: given n, compute $g(n)$, then compute $\varphi_{g(n)}(n) = \varphi_U^{(2)}(g(n), n)$, and then add one to this result.

Thus, if $\varphi_{g(n)}$ is *total* for every n we may conclude that

$$h(n) = \varphi_{g(n)}(n) + 1$$

is a total Turing-computable function. But this h cannot equal $\varphi_{g(n_0)}$ for any n_0, for then we would have the contradiction $h(n_0) = \varphi_{g(n_0)}(n_0) = \varphi_{g(n_0)}(n_0) + 1$. Thus, we conclude that for any total Turing-computable numerical function $g: \mathbf{N} \to \mathbf{N}$, the enumeration $\varphi_{g(0)}, \varphi_{g(1)}, \varphi_{g(2)}, \ldots$ cannot contain all (and only) the total functions. $\qquad\square$

We now come to the central result of this section.

3 Theorem (The Unsolvability of the Halting Problem). *Consider the function* halt: $\mathbf{N} \to \mathbf{N}$ *defined as follows*:

$$\text{halt}(x) = \begin{cases} 1, & \text{if } M_x \text{ halts on input } x; \\ 0, & \text{if } M_x \text{ never halts on } x. \end{cases}$$

Then halt *is not Turing-computable.*

Before proving this result we make a few observations. First of all, notice that *halt* is a bona fide total function. It has a definite value for every input. What we are trying to show is that *halt* is *not* computable: there is no method (formally expressed as a Turing machine) for calculating *halt*'s values. Notice also that we have been very conservative in our statement of Theorem 3. We have only claimed Turing noncomputability. In fact, if Church's thesis is subscribed to, *halt* is not computable by any algorithm specified in any programming language.

PROOF OF THE THEOREM. Suppose we construct an $\mathbf{N} \times \mathbf{N}$ array, setting the (i, j) entry to $\downarrow$ if the computation of $\varphi_i(j)$ terminates, while setting it to $\uparrow$ if not. To construct a function φ not in the set of φ_i's we adopt Cantor's diagonal argument: moving down the diagonal, we "flip the arrows," giving $\varphi(x)$ the halting behavior opposite to that of $\varphi_x(x)$:

$$\varphi(x) = \begin{cases} 1 & \text{if } \varphi_x(x) \text{ is not defined} \\ \bot & \text{if } \varphi_x(x) \text{ is defined.} \end{cases}$$

But this just says that

$$\varphi(x) = \begin{cases} 1 & \text{if } halt(x) = 0; \\ \bot & \text{if } halt(x) = 1. \end{cases}$$

Now clearly if *halt* is computable, there certainly exists a Turing machine that computes φ. However, our diagonal argument guarantees that φ is *not* computable, for if φ were φ_j for some j we would have $\varphi_j(j) = 1$ if $\varphi_j(j) = \perp$ while $\varphi_j(j) = \perp$ if $\varphi_j(j) = 1$—a contradiction. Therefore no calculation method for *halt* can exist.   □

Thus we have exhibited a natural, even interesting and valuable, general problem in computer science that has no algorithmic solution. (This result is extremely fundamental and is closely related to Gödel's famous result on the incompleteness of formal theories of arithmetic. For a discussion of this result and its relationship to the theory of computability, see Section 7.3 of *A Programming Approach to Computability*.)

At this point let us clarify our thinking about computable sets of natural numbers.

4 Definition. A set $S \subset N$ is *decidable* if membership in S can be determined algorithmically. That is, S is decidable (or *recursive*, or *solvable*) if the characteristic function of S,

$$\chi_S(x) = \begin{cases} 1 & \text{if } x \in S \\ 0 & \text{if } x \notin S \end{cases}$$

is Turing-computable.

If S is not decidable, we say S is *undecidable*, or *unsolvable*, or *nonrecursive* or, in cases requiring considerable emphasis, *recursively unsolvable*.

So far we have one example of an undecidable set: the set

$$K = \{n \,|\, \varphi_n(n) \text{ returns a value}\}$$

is recursively unsolvable. Our next result shows how to embed the undecidability of K in language theory, using as our embedding technique the proof of Theorem 5.3.4.

5 Definition. Let G be a Type 0 grammar over some alphabet $V \cup X$. The *deduction problem* for G is the problem of determining, given arbitrary strings x, y in $(V \cup X)^*$, whether $x \overset{*}{\Rightarrow} y$ in G.

6 Theorem. *The deduction problem for Type 0 grammars is undecidable.*

PROOF. The proof is based on the following reduction. We show that the question $S \overset{*}{\Rightarrow} w$ in G, for G an arbitrary Type 0 grammar, w an arbitrary string from X^*, is undecidable. Clearly, if no algorithm exists for this problem, then none can exist for the more general problem $x \overset{*}{\Rightarrow} y$ in G for x, y in $(V \cup X)^*$. But in the proof of Theorem 5.3.4 we showed how to go from an arbitrary Turing machine M to a grammar G such that $S \overset{*}{\Rightarrow} w$ in G if and only if M, when started on a tape initialized to w, eventually halts. Thus if we consider the grammatical encoding of M with tape initialized to the unary encoding of

M's index, then a decision procedure—an algorithm—for settling questions of the form "does $S \overset{*}{\Rightarrow} w$?" would, in effect, be a decision procedure for the halting problem. But such an algorithm cannot exist, by Theorem 3. □

By contrast to the previous result, we have the following:

7 Theorem. *The deduction problem for context-sensitive grammars is decidable.*

PROOF. Let G be a context-sensitive grammar written in the form in which productions never decrease length. As a result, if x and y are sentential forms and $|y| < |x|$ then $x \overset{*}{\Rightarrow} y$ is clearly impossible. If $|x| \leq |y|$, then there are only finitely many sentential forms of a length k, with $|x| \leq k \leq |y|$, say n. Apply all possible production rules of G to x in all possible orders. If a sentential form which has been generated previously is produced, abandon processing. Either y is eventually reached, or only strings are generated which are longer than y. In the latter case, y is not reachable from x. □

8 Corollary. *The deduction problem for context-free grammars is decidable.*

PROOF. Immediate from Theorem 7, since every context-free language is context-sensitive. □

9 Definition. Given a class of grammars C, the *emptiness problem* for C is the problem of determining for arbitrary G in C, if the language $L(G)$ is empty.

10 Theorem. *The emptiness problem for context-free grammars is decidable.*

PROOF. Exercise 3. □

We now turn to "Post's correspondence problem," a classical undecidability result about string equations, discovered by Emil L. Post, with direct applications to language theory.

11 Definition. A *Post system* P over a finite alphabet X is a set of ordered pairs (y_i, z_i), $1 \leq i \leq n$, where y_i, z_i are strings in X^*. A pair (y_i, z_i) is sometimes called a *Post equation*. The *Post correspondence problem* (PCP) is the problem of determining, for an arbitrary Post system P, whether there exist integers $i_1, \ldots, i_k$ such that

$$y_{i_1} y_{i_2} \cdots y_{i_k} = z_{i_1} z_{i_2} \cdots z_{i_k}$$

The i_j's need not be distinct. For a given PCP, we shall refer to a solution string as a *Post string*.

12 Example. Consider the Post system over $X = \{0, 1\}$:

$$P = \{(1, 01), (\lambda, 11), (0101, 1), (111, \lambda)\}$$

This system has a solution, since

$$y_2 y_1 y_1 y_3 y_2 y_4 = 110101111 = z_2 z_1 z_1 z_3 z_2 z_4.$$

The following theorem is a fundamental result.

13 Theorem. *The Post correspondence problem is unsolvable.*

PROOF. We reduce the deduction problem for Type 0 grammars to the PCP. That is, we show that given any instance of the deduction problem for Type 0 grammars, there is a corresponding Post system with the property that the deduction problem has a solution if and only if the corresponding Post system has a solution. Then a decision procedure for PCP could be used to solve instances of the deduction problem. But we know that this problem is unsolvable by Theorem 6.

We proceed with the proof as follows. Given an arbitrary grammar G over alphabet $\hat{V} = V \cup X$ we add new symbols to $\hat{V}$ as follows, and call the resulting alphabet V':

(1) For every a in $\hat{V}$, add a new symbol a';
(2) Add new symbols $*$, $*'$, [, and] to V'. If string w is a string in $\hat{V}$, we write w' for the string constructed by replacing every symbol of w by its primed companion.

Given a deduction problem $x \overset{*}{\Rightarrow} y$ in G, we transform it into a Post correspondence problem for a Post system P which is suitable for encoding grammatical derivations of the form

$$w_1 \Rightarrow w_2 \Rightarrow \cdots \Rightarrow w_n$$

with $x = w_1$ and $y = w_n$. The terms of the Post equations for such a sequence will have a special form.

To simplify comparison between strings over $\hat{V}^*$, we shall now write Post equations in the Post system P in the form $\frac{y_i}{z_i}$ rather than (y_i, z_i). Our task is to define a set of $\frac{y_i}{z_i}$ such that we can find indexes $i_1, i_2, \ldots, i_k$ for which we have that

$$y_{i_1} y_{i_2} \cdots y_{i_k}$$

$$z_{i_1} z_{i_2} \cdots z_{i_k}$$

are two ways of encoding the *same* legal derivation $x \Rightarrow w_2 \Rightarrow \cdots \Rightarrow y$ of G. We shall build up our Post string so that alternating intermediate forms and associated asterisks are primed:

$$[x * w_2' *' w_3 * w_4' *' \ldots y]$$

Indeed we start from a legal derivation $x \Rightarrow w_2 \Rightarrow \cdots \Rightarrow y$ and give a sequence of equations from P such that the preceding string is a solution of P.

We first include the pair $\begin{smallmatrix}[x*\\[\end{smallmatrix}$ in the Post system to initiate this string.

Since $x \Rightarrow w_2$, we can write $x = a_1 \ldots a_k u b_1 \ldots b_l$ and $w_2 = a_1 \ldots a_k v b_1 \ldots b_l$ with $u \to v$ a production of G. We will take this production into account in our Post system by adding equations which will build the sequence

$$a_1' \ldots a_k' v' b_1' \ldots b_l' *'$$

$$a_1 \ldots a_k u b_1 \ldots b_l *$$

These equations will allow us to extend

$$[x *$$
$$[$$

to

$$[x * w_2' *'$$
$$[x *$$

Since $w_2 \Rightarrow w_3$ we can similarly write $w_2 = c_1 \ldots c_r u_1 d_1 \ldots d_s$ and $w_3 = c_1 \ldots c_r v_1 d_1 \ldots d_s$ with $u_1 \to v_1$ a production of G. Again, we introduce equations that allow us to form the following Post sequences:

$$c_1 \ldots c_r v_1 d_1 \ldots d_s *$$

$$c_1' \ldots c_r' u_1' d_1' \ldots d_s' *'$$

Using these equations we extend

$$[x * w_2' *' \qquad [x * w_2' *' w_3 *$$
$$[x * \qquad\text{to}\qquad [x * w_2' *'$$

We may repeat this process. In fact, if P contains

(a) The pair $\begin{smallmatrix}[x*\\[\end{smallmatrix}$;
(b) The pairs $\tfrac{a'}{a}$ and $\tfrac{a}{a'}$ for each a in $\hat{V} \cup \{*\}$; and
(c) The pairs $\tfrac{v'}{u}$ and $\tfrac{v}{u'}$ for each production $u \to v$ of G,

then it is clear that for any valid derivation $x \Rightarrow w_2 \Rightarrow w_3 \Rightarrow \cdots$ we can concatenate a suitable sequence of pairs to obtain both

$$[x * w_2' *' w_3 \cdots *' w_{2k-1} * w_{2k}' *'$$
$$[x * w_2' *' w_3 * \cdots *' w_{2k-1} * \qquad\qquad \dagger$$

and

$$[x * w_2' *' w_3 * \cdots * w_{2k}' *' w_{2k+1} *$$
$$[x * w_2' *' w_3 * \cdots * w_{2k}' *' \qquad\qquad \dagger\dagger$$

for each k. Now suppose the derivation is such that $w_{2k}' = y'$, where y is the concluding string of our sought for derivation $x \overset{*}{\Rightarrow} y$. Then the pair $\begin{smallmatrix}]\\y*]\end{smallmatrix}$ will complete the pairs of (†) to solve the correspondence problem for P. Similarly, if $w_{2k+1} = y$, the pair $\begin{smallmatrix}]\\y*]\end{smallmatrix}$ will complete a solution. In other words, to complete the specification of P we add to (a), (b), (c) above the pairs:

(d) $_{y*}{}^]$ and $_{y'*}{}^]$

We conclude that "if $x \overset{*}{\Rightarrow} y$ in G then P has a solution to its correspondence problem." (Note that in the trivial case $x = y$, the solution to P is simply obtained by concatenating the pairs $[^{x*}$ and $_{y*}]$.)

To complete the proof we need to show that if P as constructed previously has a solution, then $x \overset{*}{\Rightarrow} y$ in G. We leave the proof of this claim to the reader as an exercise.   □

Because it is equation-oriented, PCP is particularly well suited for applications in language theory.

14 Theorem. *The problem of deciding whether an arbitrary context-free grammar is ambiguous is undecidable.*

PROOF. Let $P = \{(x_1, y_1), \ldots, (x_n, y_n)\}$ be a Post system over X. Augment X with n new symbols $\bar{1}, \bar{2}, \ldots, \bar{n}$, and consider the following context-free grammar:

$$S \to S_1 | S_2$$

$$S_1 \to x_1 S_1 \bar{1} | \ldots | x_n S_1 \bar{n} | x_1 \bar{1} | \ldots | x_n \bar{n}$$

$$S_2 \to y_1 S_2 \bar{1} | \ldots | y_n S_2 \bar{n} | y_1 \bar{1} | \ldots | y_n \bar{n}$$

This grammar is ambiguous if and only if the corresponding Post problem P has a solution. This is because S_1 can only derive strings of the form $x_{i_1} x_{i_2} \ldots x_{i_n} \bar{i}_n \ldots \bar{i}_2 \bar{i}_1$ and S_2 can only derive strings of the form $y_{i_1} y_{i_2} \ldots y_{i_n} \bar{i}_n \ldots \bar{i}_2 \bar{i}_1$. Clearly, ambiguity only arises if a string derived from S_1 equals a string derived from S_2. But such equality then encodes a solution to the correspondence problem for P.   □

EXERCISES FOR SECTION 5.4

1. Explain (in the spirit of the alphabet lemma of the last section) why a two-letter tape alphabet is sufficient to perform any Turing machine computation.

2. Assuming that V is fixed, and that X is a finite subset of $\{v_0, v_1, v_2, v_3, \ldots\}$, give a coding for triples (x, y, G) where x and y belong to $(V \cup X)^*$, and G is a type 0 grammar. Do this so that if D_k is the kth triple, then the triple itself is algorithmically recoverable from k; and given any triple its code in $\mathbf{N}$ is algorithmically producible. (Outline the algorithm. You don't need to specify a Turing machine formally.)

3. Prove Theorem 10. (*Hint*: Use the proof of the pumping lemma.)

4. Is the problem "Is $L(G)$ finite?" decidable for arbitrary context-free G?

5. Is the Post system $P = \{(101, 10), (11111, 11), (\lambda, 1)\}$ solvable?

6. Prove that a Post system over a one-symbol alphabet is decidable.

7. Give a systematic listing of all Post systems over the alphabet $\{a, b\}$.

8. Complete the proof of Theorem 13.

9. Show that there exists a Turing machine M such that $K = T(M)$.

10. This exercise establishes the following result: there exist context-free languages L_1 and L_2 and a homomorphism h such that $L = h(L_1 \cap L_2)$ is undecidable. For the purposes of this exercise define a *computation* of a Turing machine M to be a sequence

$$w_1 * w_2^R * w_3 * w_4^R \cdots * w_n$$

such that

(1) each w_i is an ID of M, $1 \le i \le n$
(2) w_1 is an initial ID of M
(3) w_n is a final ID of M
(4) $w_i \Rightarrow w_{i+1}$ for $1 \le i < n$

   (a) For fixed Turing machine M, show that $\{w * z^R | w \Rightarrow z\}$ and also $\{w^R * z | w \Rightarrow z\}$ are context-free languages.
   (b) Construct L_1 and L_2 so that $L_1 \cap L_2$ represents the set of computations of M. (*Hint*: Construct L_1 and L_2 in the form of symbols separated into blocks by "*"'s. Construct L_1 so that if i is odd, $w_i \Rightarrow w_{i+1}^R$. Construct L_2 so that if i is even, $w_i^R \Rightarrow w_{i+1}$. Use (a) to establish that L_1 and L_2 are context-free.)
   (c) Modify the construction in (b) slightly, and propose a Turing machine M and homomorphism h for which $h(L_1 \cap L_2)$ is an undecidable set. (*Hint*: Use Exercise 9.)

CHAPTER 6

Fixed Point Principles in Language Theory

6.1 Partial Orders and Fixed Points
6.2 Fixed Point Theory and Context-Free Languages

Stir a cup of coffee with a spoon. It is a well-known fact of popular mathematics that some point in the coffee will be in its initial position after the stir. This drop of coffee is called a *fixed point*. Fixed points are extremely important objects in mathematics and computer science. In numerical methods, fixed points can be used to find roots of equations; in programming semantics, fixed point methods can be used to find the semantics (the meaning) of a program. Fixed point methods are useful in language theory too.

The basic idea can be seen by looking at a context-free grammar G for the language of matched parentheses.

$$S \to SS$$

$$S \to (S)$$

$$S \to ()$$

As an alternate way to describe $L(G)$, suppose we write the following equation:

$$g(L) = L \cdot L + (L) + ()$$

Here we have used regular set notation (Section 2.3) to abbreviate the set equation

$$g(L) = \{W_1 \cdot W_2 | w_1, w_2 \in L\} \cup \{(w)|w \in L\} \cup \{()\}$$

If we calculate $g(\varnothing)$, we obtain $\varnothing \cdot \varnothing + (\varnothing) + () = \{()\}$. Moreover, $g(g(\varnothing)) = g(\{()\}) = \{()(), (()), ()\}$. Notice: $g(\varnothing)$ and $gg(\varnothing)$ are subsets of the language generated by G, and $\varnothing \subset g(\varnothing) \subset gg(\varnothing)$. The results of this chapter will show that this is no accident: It will turn out that for all n, $g^n(\varnothing) \subset$

$g^{n+1}(\emptyset), g^n(\emptyset) \subset L(G)$, and $\bigcup_n g^n(\emptyset) = L(G)$. Indeed, we can write

$$L(G) = g(L(G))$$

We call $L(G)$ a *fixed point* of g because $L(G)$ is left unchanged when g is applied to it. The central result of Section 6.2 will be that $L(G)$ is the *least* fixed point of g; i.e., $L(G)$ is the smallest set L for which $L = g(L)$. But first, before considering fixed point methods in language theory, we devote Section 6.1 to more general properties of fixed points for functions defined on arbitrary partially ordered sets.

6.1 Partial Orders and Fixed Points

In this section we set the stage for a discussion of fixed point methods in language theory by considering partially ordered sets, a class of objects which includes many important mathematical structures such as the real numbers, the natural numbers and the set of all subsets of a nonempty set. A special collection of partially ordered sets, which we call *complete partially ordered* (cpo) sets, is particularly important for our theory of fixed points because members of the class admit useful "limit" methods.

1 Definition. A *partially ordered set* $(P, \leq)$ (*poset* for short) is a set P equipped with a binary relation $\leq$, called the *ordering relation*, such that for all x, y, z in P we have

(Reflexivity) $x \leq x$

(Transitivity) if $x \leq y$ and $y \leq z$ then $x \leq z$

(Antisymmetry) if $x \leq y$ and $y \leq x$ then $x = y$

Notice that a poset P is in general an abstract set (and not necessarily a set of numbers, say), and that $\leq$ has no a priori meaning. Rather, the binary relation $\leq$ must be specified when a partially ordered set is proposed, and that binary relation must satisfy the three axioms above to be admissible.

2 Example. Let P be the set of real numbers and let "$x \leq y$" have its usual meaning: "x is less than or equal to y." Then $(P, \leq)$ is a poset. However, note that $(P, \leq)$ would still be a poset even if we gave "$x \leq y$" the *opposite* meaning that "x is greater than or equal to y."

If $(P, \leq)$ is any poset and if A is any subset of P, then the *restriction* of $\leq$ to A (also denoted $\leq$; so for a, b in A, $a \leq b$ in the A-sense if and only if $a \leq b$ in the P-sense) obviously continues to satisfy the three axioms. Thus $(A, \leq)$ is a poset. We sum this up by saying "any subset of a poset is a poset." Thus any subset of the set of real numbers, e.g., the integers, the natural numbers, the rationals, the positive reals, or the Cantor middle-third set equipped with the less-than-or-equal-to ordering, is an example of a poset.

The symbol $\leq$ has some special properties that allow the following notational conventions in a poset $(P, \leq)$:

$$x < y \text{ if } x \leq y \text{ but } x \neq y$$

$$x > y \text{ if } y < x$$

$$x \geq y \text{ if } y \leq x$$

If x and y are real numbers then at least one of $x \leq y$ or $y \leq x$ holds. This is not the case in the following example, which is why one says *partial* order in general.

3 Example. Let A be any set and let P be the set 2^A of all subsets of A. Define the ordering relation to be the relation $\subset$ ("subset of"). Then $(P, \subset)$ is a poset: surely if X, Y, Z are subsets of A then

$$X \subset X;$$

$$X \subset Y \text{ and } Y \subset Z \text{ implies } X \subset Z; \text{ and}$$

$$\text{if } X \subset Y \text{ and } Y \subset X \text{ then } X = Y.$$

However (unless A is the empty set or is a one-element set), the poset $(P, \subset)$ has *incomparable* elements—elements for which neither $X \subset Y$ nor $Y \subset X$ holds.

4 Example. Let A be a nonempty set, and define $a \leq b$, where a and b are in A, to hold exactly when $a = b$. In this poset, all distinct elements are incomparable. This ordering is called the *discrete ordering*. (In other words, the conditions of Definition 1 are satisfied even though it does not place the elements of A in any "order" in the colloquial sense of the word.)

We next give several definitions which will refine our understanding of posets.

5 Definition. Let $(P, \leq)$ be a poset and let A be any subset of P. A *least element* of A is an element a in A satisfying $a \leq b$ for all b in A.

If a and b were both least elements of A then we would have $a \leq b$ and $b \leq a$ so that $a = b$ by antisymmetry. Thus A has either no least element or exactly one. Note, for example, that the set $\mathbf{N}$ of natural numbers has a least element, 0, but the set $\mathbf{Z}$ of all integers, positive, negative, or zero, has none. Note also that the empty set, $\varnothing$, is the least element of any poset based on the subset containment operator $\subset$ as described in Example 3.

6 Definition. Let $(P, \leq)$ be a poset and let S be any subset of P. An *upper bound* of S is an element x of P (not necessarily in S) satisfying the condition "for all $s \in S$, $s \leq x$." Let $UB(S)$ denote the set of all upper bounds for S. (Notice that $UB(\emptyset) = P$ since if $x \in P$ the condition "for all $s \in \emptyset$, $s \leq x$" is vacuously true.) The *least upper bound* for S, denoted by $LUB(S)$ or $\bigvee S$, is the least element of $UB(S)$. $LUB(S)$ may not exist, but if it does it is the unique x in P such that

(a) $s \leq x$ for all $s \in S$

(b) $x \leq y$ for all $y \in UB(S)$

7 Observation. For every set A, any collection S of elements of 2^A (i.e., subset of A) has a least upper bound, namely, $LUB(S) = \bigcup S$. To establish this claim we must first prove that the union is an upper bound for S. Then we must prove that the union is the least upper bound. The first claim is easy: for all $X \in S$, we have $X \subset \bigcup S$, so that $\bigcup S$ is an upper bound of S. Now suppose, by way of contradiction, that $\bigcup S$ was not the *least* upper bound. Then there must exist a set $Y \in UB(S)$ such that $\bigcup S \subset Y$ is false. It follows that there would be an $x \in \bigcup S$ which was not in Y. But since $x \in \bigcup S$, $x \in Z$ for some Z in the collection S, and then Z cannot be a subset of Y—contradicting Y's claim to be in $UB(S)$.

If $(x_n | n = 0, 1, 2, \ldots)$ is a sequence of elements in P, we write $LUB(x_n)$ or $\bigvee_{n \geq 0} x_n$ to denote the *least* upper bound of the set $\{x_n | n = 0, 1, 2, \ldots\}$. We sometimes abbreviate $\{x_n | n = 0, 1, 2, \ldots\}$ by writing (x_n).

A *chain* in a poset $(P, \leq)$ is a sequence (x_n) of elements in P such that $x_n \leq x_{n+1}$ for $n = 0, 1, 2, \ldots$.*

8 Definition. A poset $(P, \leq)$ is a *complete partially ordered* set (abbreviated (cpo) if it has a least element $\bot$ (read "bottom"), and if whenever (x_n) is a chain in $(P, \leq)$ then $LUB(x_n)$ exists as an element of P.

9 Example. In the poset $(\mathbf{N}, \leq)$ of the natural numbers, every finite chain $(x_0, x_1, \ldots, x_k)$ or every eventually constant chain

$$x_0 \leq x_1 \leq x_2 \leq \cdots \leq x_k = x_{k+1} = x_n \qquad \text{for all } n \geq k$$

has its maximal element for least upper bound, but a chain with no maximal element, e.g., $x_k = 2^k$, has no upper bound. If, however, we add to $\mathbf{N}$ the element ∞ and extend $\leq$ by decreeing that $x \leq \infty$ for all $x \in \mathbf{N}$ we see that

$$\bigvee \{x_n\} = \infty \text{ if } \{x_n\} \text{ has no maximal element}$$

so that $(\mathbf{N} \cup \{\infty\}, \leq)$ is a complete partially ordered set with least element 0.

10 Example. In the poset of real numbers, there is no least element and $\{x | 0 \leq x\}$ has no upper bounds. The real numbers do have an important

---

* Strictly, this is an *ascending* chain, but since we shall not consider descending chains here, we omit the adjective.

completeness property: if (x_n) is a chain of real numbers such that $UB(x_n)$ is nonempty, then $LUB(x_n)$ exists. On the other hand, if P is the set of real numbers $\{x|x < 0\} \cup \{x|0 < x\}$ then $(P, \leq)$ is a poset, but we can define a chain (x_n):

$$x_0 = -1, x_1 = -\tfrac{1}{2}, \ldots, x_n = -1/(n+1), \ldots$$

for which $UB(x_n) = \{x|0 < x\}$ is nonempty but $LUB(x_n)$ does not exist, as there is no smallest number in P strictly greater than 0.

11 Example. If $(\mathbf{N}, \leq)$ is the set of natural numbers $\{0, 1, 2, \ldots\}$, then any nonempty subset of $\mathbf{N}$ has a least element. This property (sometimes called the well-ordering principle) is equivalent to the principle of mathematical induction (Exercise 7).

12 Example (Partial Orderings on X^*). Let X be a finite alphabet, and X^* the set of all finite-length strings over X. We can impose various partial orderings on X^*. We give a sample in this example. To distinguish between these orderings we subscript the ordering relation symbol $\leq$.

Suppose $X = \{a_1, a_2, \ldots, a_k\}$. We obtain a *lexicographic* order $\leq_{lex}$ on X^* by first imposing an ordering on the symbols of X, say $a_1 \leq_{lex} a_2 \leq_{lex} \cdots \leq_{lex} a_k$. Then for all $x = x_1 \ldots x_m$ and $y = y_1 \ldots y_n$, where $m, n \geq 0$ and $x_i, y_j \in X$, we set $x \leq_{lex} y$ if and only if:

(1) $x_i = y_i$ for $i = 1, \ldots, m$ and $m \leq n$; or
(2) there is a $j \geq 1$ such that $x_i = y_i$ for $i = 1, \ldots, j-1$ and $x_j <_{lex} y_j$.

If $X = \{a, b, c, \ldots, z\}$, the Roman alphabet in its usual order, then the lexicographic order on X^* is the familiar "alphabetical order" used in dictionaries and telephone books.

We next define the *prefix* order, the *suffix* order, and the *substring* order on X^*, denoted by $\leq_{pre}$, $\leq_{suf}$, and $\leq_{sub}$, respectively. For all x and y in X^* we say that:

$x \leq_{pre} y$ iff there is a $w \in X^*$ such that $xw = y$ (x is a prefix of y);

$x \leq_{suf} y$ iff there is a $v \in X^*$ such that $vx = y$ (x is a suffix of y);

$x \leq_{sub} y$ iff there are $v, w \in X^*$ such that $vxw = y$ (x is a substring of y).

$\leq_{pre}$, $\leq_{suf}$, and $\leq_{sub}$ are partial orderings of X^* (check it!). Note that $\leq_{sub}$ *refines* both $\leq_{pre}$ and $\leq_{suf}$, in the sense that if $x \leq_{pre} y$ or $x \leq_{suf} y$ then $x \leq_{sub} y$.

We define another partial ordering on X^* which is often encountered in studies of computer science. For all x and y in X^* let us say that x is *injected* in y, denoted $x \leq_{inj} y$, iff $x = x_1 \ldots x_m$ and $y = y_1 x_1 \ldots y_m x_m y_{m+1}$ for some $m \geq 0$ and $x_i, y_j \in X^*$ where $i = 1, \ldots, m$ and $j = 1, \ldots, m+1$. We leave it to the reader to verify that $\leq_{inj}$ is indeed a partial ordering of X^* (Exercise 4).

With respect to each of $\leq_{lex}$, $\leq_{pre}$, $\leq_{suf}$, $\leq_{sub}$, and $\leq_{inj}$, X^* has a least element, namely, the empty string λ. However, with respect to none of them is X^* a cpo (why?).

For a continuation of the preceding discussion, the reader is referred to Exercise 5.

The next example is crucial to the study of fixed point principles in language theory and should be studied with great care.

13 Example. Let us consider again the poset $(2^A, \subset)$ of Example 3. This structure is often studied as a Boolean algebra with meet, join, and complement operators: $\cap$ (intersection), $\cup$ (union), and $^-$ (complement), respectively. It is clear from Observation 7 that this poset is a cpo.

The least element is the empty set $\emptyset$, and if (X_n) is a chain of subsets of A, then $LUB(X_n) = \bigcup X_n$, which always exists as a subset of A.

We generalize this poset by considering k-tuples of subsets of A, where k is a fixed positive integer. Such a k-tuple is formally a member of $(2^A)^k = 2^A \times 2^A \times \cdots \times 2^A$ (k times). The poset made up of such objects is ordered as follows: For all Y and Z in $(2^A)^k$, $Y \leq Z$ if and only if $[Y]_i \subset [Z]_i$ for $i = 1, 2, \ldots, k$, where $[Y]_i$ denotes the ith component of a k-tuple Y.

The poset $((2^A)^k, \leq)$ is a cpo. Its least element is the k-tuple $(\emptyset, \emptyset, \ldots, \emptyset)$. And if (X_n) is a chain of k-tuples in $(2^A)^k$ then

$$LUB(X_n) = (\bigcup [X_n]_1, \bigcup [X_n]_2, \ldots, \bigcup [X_n]_k)$$

which always exists as a member of $(2^A)^k$.

In language theory we will be particularly interested in the case $A = X^*$. This is because, as we shall see in the next section, a context-free grammar can be thought of as a mapping from tuples of sets to tuples of sets such that the generated language is the least fixed point of the mapping. Before we turn to this analysis, however, we develop a basic result, Kleene's fixed point theorem, which will give us some general machinery for reasoning about fixed points. (This is the same Kleene, but not the same theorem, that we encountered in Section 2.3.) Roughly speaking, the fixed point theorem says that if $(D, \leq)$ is a cpo, then any suitably well-behaved map $f: D \to D$ must have a fixed point. That is, there must be some element d in D such that $f(d) = d$.

14 Definition. Let $(A, \leq)$ and $(B, \leq)$ be posets.* A function $f: A \to B$ is *monotone* from $(A, \leq)$ to $(B, \leq)$ if it is order-preserving, i.e., if $x \leq y$ implies $f(x) \leq f(y)$.

15 Example. The map from $(\mathbf{R}, \leq)$ to $(\mathbf{R}, \leq)$ that sends real number x to real number $x + 1$ is clearly monotone.

---

* We often use $\leq$ to indicate two different partial orders, one on A and one on B, where no ambiguity can result. If special care is required, we can use some special notation, such as $\leq_A$ and $\leq_B$.

16 Example. Let A and B be non-empty sets such that $A \subset B$ and $b \in B - A$. Then the map $f = 2^A \to 2^B$ that sends $C \in 2^A$ to $C \cup \{b\} \in 2^B$ is monotone.

17 Example. Consider the partial orderings on X^* defined in Example 12. Given a fixed $a \in X$, the map which sends an arbitrary $x \in X^*$ to ax is monotone on $(X^*, \leq_{lex})$, i.e., from $(X^*, \leq_{lex})$ to $(X^*, \leq_{lex})$, and on $(X^*, \leq_{pre})$, while the map which sends x to xa is monotone on $(X^*, \leq_{suf})$. The map which sends an arbitrary x to its reverse x^R is monotone from $(X^*, \leq_{pre})$ to $(X^*, \leq_{suf})$.

18 Example. Consider the poset $(2^A, \leq)$ of Example 3, where now $A = X^*$, the set of all finite length strings over the alphabet X. For an arbitrary language $L \subset X^*$, define

$$Pre(L) = \{x \in X^* | \text{there is a } y \in L \text{ such that } x \leq_{pre} y\}.$$

That is, $Pre(L)$ is the set of all prefixes of L.

Likewise, define $Suf(L)$, $Sub(L)$, $Inj(L)$, and $Lex(L)$—by respectively substituting $\leq_{suf}$, $\leq_{sub}$, $\leq_{inj}$, and $\leq_{lex}$ for $\leq_{pre}$ in the definition of $Pre(L)$. The maps which send L to $Pre(L)$, L to $Suf(L)$, L to $Sub(L)$, L to $Inj(L)$, and L to $Lex(L)$, are all monotone on $(2^A, \leq)$.

The next definition is especially interesting because it ties discrete mathematics to the concept of limits, an idea which is usually seen only in continuous mathematics.

19 Definition. Let $(D, \leq)$ and $(D', \leq)$ be cpo's. A map $f: D \to D'$ is *continuous* from $(D, \leq)$ to $(D', \leq)$ if, for any chain $x_1 \leq x_2 \leq x_3 \leq \cdots$ in D, the sequence $\{f(x_n) | n \geq 0\}$ in D' has a least upper bound which f "respects":

$$\bigvee f(x_n) = f(\bigvee x_n).$$

This notion of continuity agrees with the notion first seen in an elementary calculus course with $\bigvee$ taking the place of the operator *lim*. In that setting, a function f on the reals is continuous if for any sequence (x_n) of real numbers x_n with a limit $\lim_{n \to \infty} \{x_n\}$, we have that the sequence $(f(x_n))$ also has a limit, and

$$\lim_{n \to \infty} \{f(x_n)\} = f\left(\lim_{n \to \infty} \{x_n\}\right).$$

20 Lemma. *Every continuous map is monotone.*

PROOF. Given continuous $f: D \to D'$ and $a \leq b$ in D we must prove that $f(a) \leq f(b)$ in D'. Consider, then, the ascending chain

$$a \leq b \leq b \leq \cdots$$

Clearly it has least upper bound b. But then, by the definition of continuity, $\{f(a), f(b)\}$ has least upper bound $f(b)$ in D'; in other words, $f(a) \leq f(b)$. □

Note, in particular, that if f is continuous and (x_n) is an ascending chain in D, then $(f(x_n))$ will be an ascending chain in D'.

Next we formally define a fixed point.

21 Definition. Let $(A, \leq)$ be a poset, and let $f : A \to A$ be a map of A into itself. We say a in A is a *fixed point* of f if $f(a) = a$. The element a is the *least fixed point* of f if a is not only a fixed point of f but is least among all fixed points.

22 Examples. In general, a least fixed point may or may not exist. The identity map $id: \mathbf{N} \to \mathbf{N}$ has a least fixed point, namely 0, while every n in $\mathbf{N}$ is a fixed point of id. On the other hand, the successor map $s: \mathbf{N} \to \mathbf{N}$ given by $s(n) = n + 1$ has no fixed points, and the identity function on the set $\mathbf{Z}$ of all integers has no least fixed point.

23 The Interleaving Lemma. *Let* $(P, \leq)$ *be a cpo, and let* $x_1 \leq x_2 \leq x_3 \ldots$ *and* $y_1 \leq y_2 \leq y_3 \ldots$ *be chains of* P *such that*

(1) *for every* x_i *there exists a* y_j *such that* $x_i \leq y_j$; *and*
(2) *for every* y_i *there exists an* x_j *such that* $y_i \leq x_j$.

(The x_i *and* y_j *sequences are said to be* cofinal *in each other.)*
Then $\bigvee x_i = \bigvee y_j$.

PROOF. One standard way to prove that two poset elements a and b are equal is to show that $a \leq b$ and $b \leq a$. So suppose $a = \bigvee x_i$, $b = \bigvee y_j$. Now $a \geq b$ since $a \geq y_j$ for all j by the following reasoning: Pick an arbitrary y_j. By hypothesis there is a k such that $y_j \leq x_k$. Since $x_k \leq \bigvee x_j = a$, $y_j \leq a$ by transitivity. Now since y_j is arbitrary, a is an upper bound for all y_j's. But b is the least upper bound of the y_j's, and so $b \leq a$. By reversing the roles of a and b and arguing exactly as before, we have that $a \leq b$. Thus $a = b$, by antisymmetry. □

24 Theorem (Kleene's Fixed Point Theorem). *Let* $(D, \leq)$ *be a cpo and let* f *be a continuous function from* $(D, \leq)$ *to* $(D, \leq)$. *Then* f *has a least fixed point, which is given by*

$$\bigvee_{k \in \mathbf{N}} f^k(\bot),$$

where $\bot$ *is the least element of* $(D, \leq)$ *and* f^k *denotes the composition of* f *with itself* k *times.*

PROOF. Let us write a_0, a_1, a_2, etc. for $\bot, f(\bot), ff(\bot)$, etc. Now $a_0 = \bot \leq a_1$, and so by the monotonicity of f, $a_1 \leq a_2$. Moreover it is immediate by induction that the a_i's form an ascending chain: for each n, $f^n(\perp) \le f^{n+1}(\perp)$ guarantees $f^{n+1}(\perp) \le f^{n+2}(\perp)$ by the monotonicity of f. Let c be the least upper bound of the chain $a_0, a_1, a_2 \ldots$, which must exist by the definition of a cpo. By continuity, $f(c) = f(\bigvee a_i) = \bigvee f(a_i) = \bigvee a_{i+1}$. Now the chain a_0, $a_1, a_2, \ldots$ with LUB c and the chain $a_1, a_2, a_3, \ldots$ with LUB $f(c)$ are cofinal in each other and, by the interleaving lemma, must therefore have the same least upper bound. But this just says that $c = f(c)$.

To prove that c is the *least* fixed point for f, suppose d is any other fixed point for f, $f(d) = d$. We must prove that $c \le d$. Since $a_0 = \perp \le d$, $a_1 = f(\perp) \le f(d) = d$, and furthermore $a_2 = ff(\perp) \le ff(d) = d$, etc. This shows that d is an upper bound for the sequence $a_0, a_1, \ldots$, and therefore $c \le d$, because c is the least upper bound of this sequence. Therefore, c is the least fixed point for f.   □

Kleene's fixed point theorem is the result we need in order to develop a theory of context-free languages based on fixed points. We develop this theory in the next section.

EXERCISES FOR SECTION 6.1

1. Let $(Q, \le)$ be the rational numbers under the standard ordering.
   (a) Is $(Q, \le)$ a poset? Is $(Q, \le)$ a cpo? Explain.
   (b) Let $(Q', \le)$ be the structure of (a) restricted to the closed interval $[0, 1]$. Is this set a poset? Is it a cpo? Explain.

2. Let X be a nonempty set, and let $A, B \subset X$. Say $A \le B$ if $A \supset B$ (A is a superset of B). Is $(2^X, \supset)$ a poset? Is it a cpo? Explain.

3. Consider all formulas of the propositional calculus with variables $p_1, p_2, \ldots$, and connectives $\rightarrow$ (implication) and $\neg$ (not). If A and B are formulas, say $A \le B$ iff $A \rightarrow B$ is a tautology. Is the resulting structure a poset? Does it have a least element? Is it a cpo? Explain.

4. Prove that the relation $\le_{inj}$ defined in Example 12 on X^* is a partial ordering. (*Hint*: The only nontrivial part is the antisymmetry of $\le_{inj}$. First prove that if $x \le_{inj} y$ and $y \le_{inj} x$ then x and y have the same length.)

5. Let X be a nonempty finite alphabet. An ω-string α over X is a countably infinite sequence of symbols in X, $\alpha = \alpha_0 \alpha_1 \alpha_2 \ldots$, where $\alpha_0, \alpha_1, \alpha_2, \ldots \in X$. (We call α an ω-string because we can put its successive symbols in a one-to-one correspondence with ω, another name for the set of natural numbers.) Let X^ω denote the set of all ω-strings over X, and define $X^\infty = X^* \cup X^\omega$.
   (a) Extend the orderings $\le_{lex}$, $\le_{pre}$, $\le_{suf}$, $\le_{sub}$, and $\le_{inj}$ (defined on X^* in Example 14) to X^∞.
   (b) Show that X^∞ is a cpo relative to $\le_{pre}$ but not relative to $\le_{suf}$ and $\le_{sub}$.

6. Let X be a nonempty finite alphabet, and A an arbitrary subset of X^*.
   (a) Show that the map which sends $L \subset X^*$ to $L \cdot A$ is continuous on the cpo $(2^{X^*}, \le)$.

(b) Show that the map which sends $L \subset X^*$ to $L \cup A$ is continuous on the cpo $(2^{X^*}, \leq)$.

7. A poset $(A, \leq)$ is a *linear* (or *total*) *order* if for all x and y in A, $x \leq y$ or $y \leq x$. A linear order $(A, \leq)$ is said to be *well ordered* if every nonempty subset of A has a least element.

Let $(A, \leq)$ be a linear order, and P an arbitrary property applicable to all the elements of A. If $x \in A$ has property P, we write $P(x)$. $(A, \leq)$ satisfies the *principle of mathematical induction* in case the following is true: "If for every $z \in A$ we can deduce $P(z)$ from the fact that $P(y)$ for all $y < z$, then it follows that $P(x)$ for every $x \in A$." (For convenience, $x < y$ abbreviates: $x \leq y$ and $x \neq y$.)

Let $(A, \leq)$ be a linear order. We say that there is no *infinite descending chain* in $(A, \leq)$ if there is no infinite sequence (x_n) of elements in A such that: $\ldots x_3 < x_2 < x_1 < x_0$.

Prove (a), (b), and (c):

(a) If a linear order $(A, \leq)$ is well ordered, then it satisfies the principle of mathematical induction.

(b) If a linear order $(A, \leq)$ satisfies the principle of mathematical induction, then there is no infinite descending chain in $(A, \leq)$.

(c) If there is no infinite descending chain in a linear order $(A, \leq)$, then $(A, \leq)$ is well ordered.

6.2 Fixed Point Theory and Context-Free Languages

To prepare the way for a general fixed point theory for context-free languages, we return to the question of how to derive the language of matched parentheses using a fixed point technique.

1 Example. Recall that the productions for the language of matched parentheses were $S \to SS$, $S \to (S)$, and $S \to ()$. Let S be thought of as a set variable, and consider the function definition for $g: 2^{X^*} \to 2^{X^*}$ where $X = \{(,)\}$ given by

$$g(S) = SS + (S) + ()$$

This definition of g is obtained by "adding" all the productions of the grammar. We interpret the symbol $+$ as set union, and if S is any set in 2^{X^*} then SS, (S), and $()$ are the following sets: the concatenation of S with itself; $\{(\}S\{)\}$; and the singleton set $\{()\}$. For example, if $S = \{)\},)($\}$, then $g(S) = \{))))),)))(,)()(,)()),()))(,()),()(),()\}$. We shall see later that g is continuous. Thus, its least fixed point is $\bigvee g^n(\bot)$, which in this case is made up of the union of the iterates of g on $\varnothing$:

$$g(\varnothing) = \{()\}$$

$$gg(\varnothing) = g(\{()\}) = \{()(),(()),()\}$$

etc.

In fact, it is not too difficult to show that the union of this sequence exactly equals the language of matched parentheses.

2 Example. Consider the following example of a grammar G over the terminal alphabet $X = \{a, b\}$ with three variables, v_1, v_2, and v_3. Again we forego our vertical bar notation and use the symbol " $+$ " to indicate alternative production rules.

$$v_1 \to v_2 + v_3$$

$$v_2 \to av_2b + ab$$

$$v_3 \to bv_3a + ba$$

We take v_1 as start symbol. It is easy to show, using induction, that this grammar generates the language

$$L = \{a^nb^n | n \geq 1\} \cup \{b^na^n | n \geq 1\}.$$

As in Example 1, if we now view v_1, v_2 and v_3 as variables taking their values in 2^{X^*}, the grammar yields a function, in this case $g: (2^{X^*})^3 \to (2^{X^*})^3$, which we may write as

$$g(v_1, v_2, v_3) = (v_2 + v_3, av_2b + ab, bv_3a + ba).$$

Here $av_2b + ab$ is shorthand for $\{a\}v_2\{b\} \cup \{ab\}$, etc. As we saw in Example 6.1.13, $((2^{X^*})^3, \leq)$ is a cpo with least element $\bot = (\emptyset, \emptyset, \emptyset)$ and componentwise set inclusion as partial order: $Y \leq Z$, where $Y = (Y_1, Y_2, Y_3)$ and $Z = (Z_1, Z_2, Z_3)$, if and only if $Y_1 \subset Z_1$, $Y_2 \subset Z_2$, and $Y_3 \subset Z_3$. Least upper bounds are also determined componentwise, so that the LUB of an ascending chain (X_1^i, X_2^i, X_3^i) is simply $(\bigcup_i X_1^i, \bigcup_i X_2^i, \bigcup_i X_3^i)$. The function g of this example is also continuous, as we shall shortly see. Let us apply the fixed point theorem to determine the least fixed point of this g.

$$g^0(\bot) = (\emptyset, \emptyset, \emptyset)$$

$$g^1(\bot) = (\emptyset, \{ab\}, \{ba\})$$

$$g^2(\bot) = g(\emptyset, \{ab\}, \{ba\}) = (\{ab, ba\}, \{a^2b^2, ab\}, \{b^2a^2, ba\})$$

and from this it is easy to prove, by induction on m, that

$$g^m(\emptyset, \emptyset, \emptyset) = (\{a^kb^k, b^ka^k | 0 < k < m\}, \{a^kb^k | 0 < k \leq m\}, \{b^ka^k | 0 < k \leq m\})$$

If we take the LUB of the chain $\bot \leq g(\bot) \leq g^2(\bot) \leq \cdots$ we obtain

$$\bigvee_{m \geq 0} g^m(\bot) = (L(G), \{a^jb^j | j \geq 1\}, \{b^ja^j | j \geq 1\}).$$

We see that $L(G)$ is indeed the first component of the least fixed point of g.

Our general task, then, is to show that a context-free grammar G on n variables determines a continuous mapping

$$g: (2^{X^*})^n \to (2^{X^*})^n$$

mapping n-tuples of languages—subsets of X^*—to n-tuples of languages. Each component is associated with a different variable. Since g is continuous, it has a least fixed point, and the major result of this section establishes that if $(L_1,\ldots,L_n)$ is the least fixed point, and if the first component is associated with the start symbol of the grammar, then $L_1 = L(G)$, the language generated by the grammar. At the end of this section we present two valuable applications of this point of view: we show how to construct a regular expression that solves a set of simultaneous regular "set" equations, and we give a fixed point-derived algorithm for a Greibach normal form for a context-free grammar.

First, we must formalize our association of a function with a grammar and show that it is continuous.

3 Definition. Let $G = (X, V, S, P)$ be a context-free grammar with $V = \{v_1, v_2, \ldots, v_n\}$ and $S = v_1$. The productions of P may be organized in the form

$$v_j \to P_{j1} + \cdots + P_{jk_j}$$

where $P_{j1}, \ldots, P_{jk_j}$ are strings in $(V \cup X)^*$. Then we may reinterpret this production as a function

$$g_j: (2^{X^*})^n \to 2^{X^*}$$

where $g_j(V_1,\ldots,V_n)$, for $V_i \subset 2^{X^*}$, is the language obtained from the regular expression $P_{j1} + \cdots + P_{jk_j}$ by substituting the set V_i for each occurrence of the variable v_i, $1 \le i \le n$.

The *function of the grammar G* is then the function

$$g: (2^{X^*})^n \to (2^{X^*}), (V_1,\ldots,V_n) \mapsto (g_1(V_1,\ldots,V_n),\ldots,g_n(V_1,\ldots,V_n)).$$

4 Theorem. *The function* $g: (2^{X^*})^n \to (2^{X^*})^n$ *of a context-free grammar G is continuous on the cpo* $((2^{X^*})^n, \le)$.

PROOF. The theorem follows from the following lemmas, whose proof is left to the exercises:

(i) The following maps $(2^{X^*})^n \to (2^{X^*})$ are both continuous:
   (a) $(V_1,\ldots,V_n) \mapsto V_j$ for a fixed j
   (b) $(V_1,\ldots,V_n) \mapsto w$ for w a fixed string of X^*
(ii) If $h_1, h_2: (2^{X^*})^n \to (2^{X^*})$ are both continuous, then so too are
   (a) $(V_1,\ldots,V_n) \mapsto h_1(V_1,\ldots,V_n) \cdot h_2(V_1,\ldots,V_n)$
   (b) $(V_1,\ldots,V_n) \mapsto h_1(V_1,\ldots,V_n) + h_2(V_1,\ldots,V_n)$
   Clearly, repeated use of (ia), (ib), and (iia) yields continuity of any function induced by the production $Y \to Z_1 \ldots Z_k$ for $Z_i \in (V \cup X)$. Repeated use of (iib) yields the continuity of g_j. But then the continuity of g is immediate from
(iii) A map $g: (2^{X^*})^n \to (2^{X^*})^n$ is continuous iff each component $g_j: (2^{X^*})^n \to 2^{X^*}$ is continuous. $\square$

We next prove the fundamental result of this section, which characterizes a context-free language as the first component of the least fixed point of the function representing the grammar, where the first component corresponds to the start symbol of the grammar.

5 Theorem. *Let G be a context-free grammar with associated function g. Then the first component of the least fixed point of g is the context-free language $L(G)$ generated by G.*

PROOF. First we prove, by induction on the height of a string's derivation tree, that $w \in X^*$ is derivable by a tree of height $\leq n$ rooted at v_j, where v_j is an arbitrary variable, iff w appears in the jth component of $g^n(\bot)$.

When $n = 1$, $v_j \to w$ must be a production of G, and so w must also appear as a term in the definition of g_j thus placing it in $g_j(\bot)$. Conversely, if w is in $g_j(\bot)$, then w also appears as a term in the definition of g_j, and so $v_j \to w$ must be a production of G.

Suppose now that the conjecture holds for all $k \leq n$, and suppose w is derived by a tree of height at most $n + 1$, say with root v_j. Moreover, suppose the first level of the tree is the string α corresponding to the production $v_j \to \alpha$, where α is some nonempty string of terminals and nonterminals. Then by induction, each variable in α commands a tree of height $\leq n$ and so derives a string which shows up in the corresponding component of $g^n(\bot)$. But then when $g_j(g^n(\bot))$ representing the jth component of $g^{n+1}(\bot)$ is formed, these strings are concatenated together by the term in g_j representing the $v_j \to \alpha$ production—thus placing w in $g_j(g^n(\bot))$. Clearly, the argument is reversible: if w is in $g_j(g^n(\bot))$—the jth component of $g^{n+1}(\bot)$—then w is derivable by a tree of height at most $n + 1$ rooted at v_j.

We thus conclude that a string is derivable from $v_1 = S$ iff it belongs to $g_1(g^n(\bot))$ for some $n \geq 0$. But this just says that $L(G)$ is the first component of the least fixed point $\bigcup_{n \geq 0} g^n(\bot)$ of g. $\square$

In our analysis so far, we have shown how to derive a continuous function $g(v_1, \ldots, v_n)$ from the productions of a context-free grammar G with n variables, such that $L(G)$ is exactly the first component of the least fixed point of g. We can also view the definition of $g(v_1, \ldots, v_n)$ as a system of simultaneous equations which we want to solve for $v_1, \ldots, v_n$. Thus, in Example 1, we want to solve the equation

$$S = SS + (S) + ()$$

for the "set" variable S (i.e., the possible values for S are from 2^{X^*}), and in Example 2, we want to solve the simultaneous equations

$$v_1 = v_2 + v_3$$
$$v_2 = av_2b + ab$$
$$v_3 = bv_3a + ba$$

for the set variables v_1, v_2, and v_3. (Whereas in linear algebra, n linearly independent equations in n variables have a unique solution, in our present theory the fixed point solution is not necessarily unique.)

This alternative and useful point of view is adopted in the next theorem and the next example.

6 Theorem. *Let A and B be arbitrary languages over a nonempty finite alphabet X. The equation $v = Av + B$ has as least solution the language denoted by A^*B. (In particular, if A and B are regular, so is the least fixed point A^*B.)*

PROOF. The function $g(v) = Av + B$ is continuous on the cpo $(2^{X^*}, \leq)$, by Exercise 2. The least fixed point of g is $B \cup AB \cup A^2B \cup \cdots = A^*B$, by the fixed point theorem. □

Theorem 6 is a general result that applies to any equation $v = Av + B$ not necessarily derived from the productions of a grammar. (Notice that A and B are arbitrary subsets of X^*, possibly infinite.) Theorem 6 is useful in that it gives us a closed form solution A^*B directly—not available in the case of Theorem 5, for example—and we do not have to manipulate the successive iterates of a continuous function g, $g(\perp)$, $g^2(\perp)$, etc. Notice also that a similar result holds for the analogous left-linear equation $g(v) = vA + B$, which has closed form solution BA^*.

7 Example. Consider the system of equations

$$v_1 = av_1 + bv_2 + a \qquad (1)$$

$$v_2 = av_1 + av_2 + b \qquad (2)$$

where v_1 and v_2 are variables over 2^{X^*}, and $X = \{a, b\}$. By the preceding theorem, we can write the solution to the first equation as

$$v_1 = a^*(bv_2 + a). \qquad (3)$$

If we substitute the right-hand side of (3) in (2), we get, after regrouping

$$v_2 = (aa^*b + a)v_2 + aa^*a + b \qquad (4)$$

which, again by the preceding theorem, yields the least solution.

$$v_2 = (aa^*b + a)^*(aa^*a + b). \qquad (5)$$

Substituting (5) in (3), we can also obtain the least solution for v_1, $a^*(b(aa^*b + a)^*(aa^*a + b) + a)$.

The last example provides a useful technique for certain calculations involving regular sets. This technique involves the closed form solution to the equation $v = Av + B$ as well as the principle of substitution. We apply it in the next example to find a regular expression for the set accepted by a FSA or, equivalently, the language generated by a right-linear grammar. Note that if A and B are finite sets, the equation $v = Av + B$ corresponds to the productions of a right-linear grammar G with exactly one variable v; and, by Theorem 5, the least fixed point solution of the equation is exactly $L(G)$.

8 Example. Consider the following FSA:

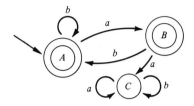

The corresponding right linear grammar (See Section 2.3) is

$$A \rightarrow bA|b|aB|a|\lambda$$
$$B \rightarrow bA|b|aC|\lambda$$
$$C \rightarrow aC|bC$$

(Remember that the production $A \rightarrow b$ signals that b sends the FSA from A to an accepting state.) We can thus write the following equations to express the relationships between the strings accepted by starting from state A and B:

$$A = bA + aB + a + b + \lambda$$
$$B = bA + b + \lambda$$

(We avoid further mention of the dead state C, since it plays no role in the solution.) One solution is obtained by substituting the second equation in the first, yielding (using $a + a = a$)

$$A = bA + abA + ab + a + b + \lambda$$

which has the least fixed point solution

$$(b + ab)^*(ab + a + b + \lambda)$$

by Theorem 6. This is a regular expression for the set accepted by the preceding FSA.

When we resume our study of parsing in the next chapter we will see a second way in which implicitly defined "regular" equations can arise.

In our most recent examples we have begun to manipulate formal language theory expressions with algebraic gusto. And indeed the algebraic laws that govern how $+$ and $\cdot$ work behave in almost exactly the way that $+$ and $\times$ work in arithmetic (the commutative law doesn't hold, though). But what about $*$? The answer is given by the following observation.

9 Observation. In arithmetic we have the following equality

$$\frac{1}{1-x} = 1 + x + x^2 + x^3 + \cdots$$

valid for any $|x| < 1$. The equation suggests that the operation * of formal language theory is the analogue of the quasi-inverse operator $x \mapsto (1-x)^{-1}$. That is, the solution

$$A*B \text{ for the set equation } S = AS + B$$

corresponds to the numerical solution

$$(1-a)^{-1}b \text{ for the equation } x = ax + b \text{ over the real numbers.}$$

The leap to algebra gives us a great deal of power, especially when we are working on some computational problem in language theory. We conclude this chapter with an elegant algebraic technique for computing a Greibach normal form for a given grammar.

10 Example (Algebraic Greibach Normal Form). Consider the following context-free grammar, which we wish to convert to Greibach normal form:

$$S \rightarrow SA + Sa + a$$

$$A \rightarrow SA + AA + aAb + b$$

Recall that in Greibach normal form, the right-hand side of each production belongs to $X \cdot V^*$. However, it is sufficient to change each production to what we may call pre–Greibach normal form (pGNF). In this normal form the right-hand side of each production belongs to $X(V \cup X)^*$. If we can change productions to pGNF, GNF follows easily because, for example, we can always replace $A \rightarrow aAb$ by $A \rightarrow aAA_1$ and $A_1 \rightarrow b$ for a suitable new variable A_1.

Using matrix notation we can rewrite the preceding grammar as

$$[S\ A] = [S\ A] \begin{bmatrix} A+a & A \\ \varnothing & A \end{bmatrix} + [a\ aAb + b] \tag{1}$$

(To check that this equation is valid, merely "multiply" it out to obtain the original grammar.)

Generalizing the technique we presented in Theorem 6, equation (1) is interesting because it decomposes the original grammar into two parts, one part with rules that are already in pre–Greibach normal form (the last term), and the matrix part, which represents the rules with a variable as leftmost symbol.

Equation (1) can be thought of as having a solution representable as

$$[S \ A] = [a \ aAb + b] \begin{bmatrix} A + a & A \\ \emptyset & A \end{bmatrix}^*$$

This looks encouraging since all the terms start with a terminal symbol, a or b. Unfortunately, the $*$ introduces an infinite set of strings. To correct this problem, note that the matrix

$$\begin{bmatrix} A + a & A \\ \emptyset & A \end{bmatrix}^+$$

is the solution of the matrix equation

$$Y = \begin{bmatrix} A + a & A \\ \emptyset & A \end{bmatrix} Y + \begin{bmatrix} A + a & A \\ \emptyset & A \end{bmatrix}$$

where

$$Y = \begin{bmatrix} Y_{11} & Y_{12} \\ Y_{21} & Y_{22} \end{bmatrix}$$

Hence equation (1) can be solved using the simultaneous equations

$$[S \ A] = [a \ aAb + b] Y + [a \ aAb + b] \tag{2}$$

$$Y = \begin{bmatrix} A + a & A \\ \emptyset & A \end{bmatrix} Y + \begin{bmatrix} A + a & A \\ \emptyset & A \end{bmatrix} \tag{3}$$

Now note that (2) is already in pGNF. Moreover, each term in (3) starts with an element of the *original* $V \cup X$ (i.e., A, S, a or b, but not one of the Y_{ij}'s). Thus if we substitute (2) into (3), we convert the latter into pGNF.

Let us first write out (2) as

$$S \rightarrow aY_{11} + aAbY_{21} + bY_{21} + a$$
$$A \rightarrow aY_{12} + aAbY_{22} + bY_{22} + aAb + b$$

Notice, however, that from (3), Y_{21} is useless, since it can never derive a terminal string. So we can simplify the last equations to

$$S \rightarrow aY_{11} + a$$
$$A \rightarrow aY_{12} + aAbY_{22} + bY_{22} + aAb + b \tag{4}$$

Next we write out the equations for (3). These are

$$Y_{11} \rightarrow aY_{11} + AY_{11} + a + A$$
$$Y_{12} \rightarrow aY_{12} + AY_{12} + AY_{22} + A$$
$$Y_{22} \rightarrow AY_{22} + A$$

If we now substitute (4) into this last equation set, then (4) and the resulting equations give us the sought-after pGNF:

$$S \to a Y_{11} + a$$

$$A \to a Y_{12} + a Ab Y_{22} + b Y_{22} + a Ab + b$$

$$Y_{11} \to a Y_{11} + a Y_{12} Y_{11} + a Ab Y_{22} Y_{11} + b Y_{22} Y_{11} + a Ab Y_{11} + b Y_{11} + a$$
$$+ a Y_{12} + a Ab Y_{22} + b Y_{22} + a Ab + b$$

$$Y_{12} \to a Y_{12} + a Y_{12} Y_{12} + a Ab Y_{22} Y_{12} + b Y_{22} Y_{12} + a Ab Y_{12} + b Y_{12}$$
$$+ a Y_{12} Y_{22} + a Ab Y_{22} Y_{22} + b Y_{22} Y_{22} + a Ab Y_{22} + b Y_{22} + a Y_{12}$$
$$+ a Ab Y_{22} + b Y_{22} + a Ab + b$$

$$Y_{22} \to a Y_{12} Y_{22} + a Ab Y_{22} Y_{22} + b Y_{22} Y_{22} + a Ab Y_{22} + b Y_{22} + a Y_{12}$$
$$+ a Ab Y_{22} + b Y_{22} + a Ab + b$$

EXERCISES FOR SECTION 6.2

1. Repeat the analysis of Examples 1 and 2 for the following context-free grammars. In each case the alphabet is $X = \{a, b\}$, and the start symbol is v or v_1.
   (a) The productions of the grammar G_1 are

   $$v \to ava | bvb | b | a | \lambda$$

   (b) The productions of the grammar G_2 are

   $$v \to ava | vbbv | a$$

   (c) The productions of the grammar G_3 are

   $$v_1 \to v_1 v_2 | v_1 v_3 | a$$
   $$v_2 \to a$$
   $$v_3 \to b$$

   (d) The productions of the grammar G_4 are

   $$v_1 \to v_2 v_3$$
   $$v_2 \to av_2 b | \lambda$$
   $$v_3 \to av_3 b | \lambda$$

2. Prove the lemmas of Theorem 4: We consider maps $g: (2^{X^*})^n \to 2^{X^*}$ where X is a nonempty finite alphabet and $k \le 1$.
   (a) Show that constant maps are continuous. A constant map sends an arbitrary $(v_1, v_2, \ldots, v_n)$ in $(2^{X^*})^n$ to some fixed $A \subset X^*$.
   (b) Show that the projection maps are continuous. A projection map sends an arbitrary $(v_1, v_2, \ldots, v_n)$ in $(2^{X^*})^n$ to v_j for some fixed j, $1 \le j \le n$.
   (c) Let g_1 and g_2 be continuous maps from $((2^{X^*})^n, \le)$ to $(2^{X^*}, \le)$. Prove that $g_1 + g_2$ and $g_1 \cdot g_2$ are both continuous, where

$$(g_1 + g_2)(v_1,\ldots,v_n) = g_1(v_1,\ldots,v_n) \cup g_2(v_1,\ldots,v_n)$$

$$(g_1 \cdot g_2)(v_1,\ldots,v_n) = g_1(v_1,\ldots,v_n) \cdot g_2(v_1,\ldots,v_n)$$

(d) Verify that the function g on $(2^{X^*})^n$ induced by a context-free grammar with n variables, as in Theorem 3, is continuous.

3. Consider the equation $v = Av + B$, under the same assumptions as in Theorem 6.
(a) Show that if $\lambda \notin A$ then the equation's least solution is also its unique solution.
(b) Show that X^* is a solution if $\lambda \in A$.

4. Find the least solution of the simultaneous equations

$$v_1 = (a + b)v_1 + (a + b)v_2 + a$$

$$v_2 = (aa)^*v_1 + (bb)^*v_2 + b$$

5. Repeat the analysis of Example 8 for the following FSA:

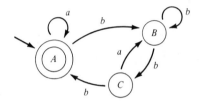

CHAPTER 7

Parsing, Part II

7.1 Top-Down Parsing and LL Grammars

In this section we consider strong *LL* grammars. This grammatical class is important because members of the class, and especially members of the class strong *LL*(1), are easy to parse and easy to write parsers for. In addition, many important programming language fragments can be given a strong *LL*(1) description, so parsing based on this grammatical characteristic has important applications in computer science.

We first recall the top-down parsing procedure which motivated our introduction of Earley's algorithm in Section 4.2. Given a grammar $G = (X, V, S, P)$ and a string $w = a_1 \ldots a_n$ over X, proceed as follows. Start by trying each of the S productions, in turn, as a plausible parse tree top. Some of these proposed starting productions may be judged incorrect immediately—an $S \rightarrow bC$ rule, for example, is obviously no good if $b \neq a_1$. Other productions may be consistent with an initial segment of w and may therefore require further analysis. With each consistent initial sentential form, attempt to build a leftmost derivation by expanding the leftmost variable in the form, again using all possible productions that match the variable to be expanded. Continue in this fashion until all partial trees are judged inconsistent, or until a successful parse is found.

In general this method will do parsing correctly, but in the worst case it will require an exponential number of rewrite rule applications (as a function of the length of the input string). In Section 4.2 we showed how to replace this process by the relatively economical process of filling in a table of dotted productions (Earley's algorithm). Here we take a different tack, asking what restrictions on grammatical form would allow the correct rule to be selected with little or no computation at each stage so that only a linear number of rewrites would be needed. In the next few examples, we investigate some grammars which have this convenient property.

1 Example. Consider the grammar

$$S \rightarrow aB | bCc | c$$

$$B \rightarrow bB | aCC$$

$$C \rightarrow aSS | c$$

Now suppose we wish to decide whether the string *aaaccc* is generated by this grammar. There are three possible starting productions, but since the string begins with an *a*, only

$$S \rightarrow aB$$

is possible. Similarly, there are two *B* productions but only the second one is possible, since the second symbol in the string is also an *a*. Thus we get

$$S \Rightarrow aB \Rightarrow aaCC$$

Applying the same principle a third time, we choose the *C* production $C \rightarrow aSS$, since this is the only way to match the third *a* in the input string:

$$S \Rightarrow aB \Rightarrow aaCC \Rightarrow aaaSSC$$

Finally, we can convert the final three symbols to *c*'s unambiguously, and we get a successful parse of the string.

Thus, this Greibach normal form grammar has the convenient property that any pair of productions with the same left-hand side can be distinguished by the right-hand lead terminal symbol.

2 Example. Consider the grammar

$$S \rightarrow aBBB | bC$$

$$B \rightarrow bB | aC$$

$$C \rightarrow SS | c$$

For this example, disambiguation of productions is not completely obvious, because the *C* productions are not in Greibach normal form. Nevertheless, since *S* must rewrite with a lead *a* or *b*, and the second *C* production is $C \rightarrow c$, the current symbol in the input stream by itself is sufficient to distinguish between the two *C* productions.

3 Example. Consider the grammar

$$S \rightarrow aBB \,|\, bC$$
$$B \rightarrow CB \,|\, aC$$
$$C \rightarrow abS \,|\, c$$

The B productions in this example at first present problems. The variable C can rewrite to a lead a, so the B productions cannot be distinguished on the basis of the next input stream symbol alone. However, if we permit ourselves to look two symbols into the input stream we can distinguish between B productions, as follows. If the first production is used, we have $B \Rightarrow CB \Rightarrow abSB$ or $B \Rightarrow CB \Rightarrow cB$ so that all strings derived must begin ab or c. If the second production is used, on the other hand, only aa or ac prefixes are possible. Therefore, a two-symbol look-ahead is adequate for "deterministic." top-down parsing in this case.

This last example informally points us toward the full notion of a strong $LL(k)$ grammar. A grammar will turn out to be strong $LL(k)$ if k-symbol look-ahead is sufficient to distinguish between the alternative productions for any variable in the grammar.

4 Definition. Let $w \in X^*$, where X is a finite alphabet. Define

$$Init_k(w) = \begin{cases} w, & \text{if } |w| \leq k \\ u, & \text{if } uv = w \text{ with } |u| = k. \end{cases}$$

Define $Init_k(L) = \{w \,|\, w = Init_k(x) \text{ and } x \in L\}$.

In other words, $Init_k(w)$ is the initial k letters of w—or all of w, if $|w| < k$. The motivating examples lead us to attempt to define $look_L(k, A \rightarrow z)$, for any production $A \rightarrow z$, as the set of terminal strings that can occur as the first k letters of a string w that can appear in any *leftmost* derivation which uses $A \rightarrow z$ as follows:

$$S \overset{*}{\underset{lm}{\Rightarrow}} uAv \underset{lm}{\Rightarrow} uzv \overset{*}{\underset{lm}{\Rightarrow}} uw$$

5 Definition. Let G be a context-free grammar and let $A \rightarrow z$ be a production of G, where $z \in (V \cup X)^*$. Then we define the *look-ahead function for leftmost derivations*

$$look_L(k, A \rightarrow z) = Init_k \{w \,|\, S \overset{*}{\underset{lm}{\Rightarrow}} uAv \underset{lm}{\Rightarrow} uzv \overset{*}{\underset{lm}{\Rightarrow}} uw \text{ with } w \in X^*\}$$

For example, $look_L(2, B \rightarrow CB) = \{ab, cc, ca\}$ in Example 3.

Using the look function, we can give a formal definition of a strong $LL(k)$ grammar.

6 Definition. A context-free grammar is *strong $LL(k)$* if, for any variable A and any productions $A \rightarrow z_1$ and $A \rightarrow z_2$, $z_1 \neq z_2$,

$$look_L(k, A \to z_1) \cap look_L(k, A \to z_2) = \varnothing$$

That is, in a strong $LL(k)$ grammar k look-ahead is sufficient to distinguish between distinct A-productions in a leftmost derivation.

It is not hard to show that the grammar in Example 3 earlier is strong $LL(2)$ but not strong $LL(1)$.

It turns out, for a variety of reasons, that strong $LL(k)$ grammars are practical for parsing when $k = 1$. When $k = 0$ the class of admissible grammars is too weak to be interesting (see Exercise 4). For $k > 1$, error recovery is complicated, and the parsing algorithm, because of the additional look-ahead required, is less efficient. Sometimes a grammar is almost strong $LL(1)$, and the $LL(1)$ algorithm outlined earlier can be used, with adjustments made for the few productions that violate the strong $LL(1)$ condition.

7 Example. The grammar for the language of **while**-programs, Example 1.4.7, which we repeat below, is strong $LL(1)$.

$$C \to \textbf{begin } S_1 \textbf{ end } \{C \text{ for compound statement}\}$$
$$S_1 \to SS_2$$
$$S \to A|W|C$$
$$A \to V := T\{A \text{ for assignment statement}\}$$
$$T \to \textbf{pred } (V)|\textbf{succ } (V)|0$$
$$S_2 \to ; S_1|\lambda$$
$$W \to \textbf{while } V \neq V \textbf{ do } S\{W \text{ for } \textbf{while} \text{ statement}\}$$
$$V \to X|Y$$

PROOF. Only the S_2 productions are a problem; it's easy to check the $LL(1)$ condition for all other productions. Before we settle the status of the S_2 productions, however, let's consider an apparent simplification to Definition 5. Suppose we use the following notion in place of $look_L$:

$$wronglook(k, A \to z) = Init_k(\{w|A \to z \overset{*}{\Rightarrow} w, w \in X^*\})$$

For the preceding grammar, we get

$$wronglook(1, S_2 \to \lambda) = \lambda$$

which gives us no useful information—all strings "start" with λ—and thus the *wronglook* concept is clearly inadequate. By contrast, since $S_2 \to \lambda$ can only be applied in the context provided by $C \to \textbf{begin } S_1 \textbf{ end}$ and $S_1 \to SS_2$ we have

$$look_L(1, S_2 \to \lambda) = \textbf{end}$$

This gives us the appropriate disambiguating information, since

$$look_L(1, S_2 \to ; S_1) = ;$$

Thus the S_2 productions satisfy the strong $LL(1)$ condition.   □

Definition 7 shows that nontrivial fragments of programming languages can be strong $LL(1)$). We are not so lucky with the next example.

8 Example. The equation grammar of Example 4.2.2 is not strong $LL(k)$ for any k.

PROOF. For any n, ($^n \in look_L(n, S \to R) \cap look_L(n, S \to (S))$).   □

Example 8 is especially interesting because we will show in the next section when we take a fresh look at bottom-up parsing (already met in our discussion of the CKY algorithm in Section 4.1) that the grammar of these equations belongs to the weakest *bottom-up* language class, the $LR(0)$ grammars.

Top-down parsing has one special weakness: the parsing style that operates by expanding the leftmost variable in a sentential form in an attempt to match the next terminal symbol in the input stream goes into an infinite loop if the grammar is left-recursive. As a formal counterpart to this observation we have the following result.

9 Theorem. Let $G = (X, V, S, P)$ be a context-free grammar with a left-recursive rule $A \to AW$, where $W \in (V \cup X)^*$. Suppose also that A can derive a terminal string, i.e., $A \overset{*}{\Rightarrow} w \in X^*$. Then G is not strong $LL(k)$ for any k.

PROOF. Exercise 3.   □

We conclude this section with a result that confirms our intuition that strong $LL(k)$ grammars are especially well-behaved grammatical systems.

10 Theorem. For fixed $k \geq 0$ let G be a strong $LL(k)$ grammar. Then G is unambiguous.

PROOF. Suppose G is ambiguous, and let

$$S \overset{*}{\underset{lm}{\Rightarrow}} uAv \underset{lm}{\Rightarrow} uz_1 v \overset{*}{\underset{lm}{\Rightarrow}} uw$$

$$S \overset{*}{\underset{lm}{\Rightarrow}} uAv \underset{lm}{\Rightarrow} uz_2 v \overset{*}{\underset{lm}{\Rightarrow}} uw$$

be two distinct leftmost derivations for uw where the application of $A \to z_1$ versus $A \to z_2$ is the step where the derivations of uw first differ. This means that the named productions have the same length k prefix of w in their look-aheads. Consequently, the grammar is not strong $LL(k)$. Moreover, since k is arbitrary the grammar is not strong $LL(k)$ for any k.   □

EXERCISES FOR SECTION 7.1

1. Is this grammar strong $LL(k)$ for any k?

$$S \rightarrow AB$$
$$S \rightarrow BC$$
$$A \rightarrow a$$
$$A \rightarrow Ba$$
$$B \rightarrow Cc$$
$$C \rightarrow c$$
$$B \rightarrow a$$

2. Complete the proof that the language of **while**-programs is strong $LL(1)$.

3. Prove Theorem 9.

4. Give an example of an infinite strictly context-free language with a strong $LL(0)$ grammar.

5. Let $k \geq 0$ be a fixed but arbitrary integer. Given an example of a context-free grammar that is not strong $LL(k)$ but is strong $LL(k + 1)$.

6. Does every regular language have a strong $LL(1)$ grammar?

7.2 Bottom-Up Parsing and LR Grammars

We now look at *bottom-up* parsing. Naive bottom-up parsing is the obvious (but not most efficient) way to parse a sentence: simply scan the sentence left-to-right and hypothesize productions that could yield the observed strings of symbols.

1 Example. Consider, as we did in Section 1.4, the context-free grammar with the following productions

$$1.\ E \rightarrow E + T$$
$$2.\ E \rightarrow T$$
$$3.\ T \rightarrow T * F$$
$$4.\ T \rightarrow F$$
$$5.\ F \rightarrow (E)$$
$$6.\ F \rightarrow a$$

$L(G)$ is the set of well-formed arithmetical expressions over $a, (,), +,$ and $*$. E stands for "expression," T for "term," and F for "factor." The grammar is in fact unambiguous, and so there is a unique rightmost derivation for each string in $L(G)$. We may try to find the *rightmost* derivation by seeking to apply productions *in reverse*, as in the following attempted parse of the string $a + a * a$:

$$a + a * a \Leftarrow F + a * a \quad \text{(reverse 6)}$$
$$\Leftarrow T + a * a \quad \text{(reverse 4)}$$
$$\Leftarrow E + a * a \quad \text{(reverse 2)}$$
$$\Leftarrow E + T * a \quad \text{(reverse 4)}$$

But now note: If we try to reduce $E + T * a$ to either $E * a$ (reverse 1) or $E + E * a$ (reverse 2) we get stuck because no right-hand side of a production contains "$E *$." This illustrates the nondeterminism (already seen in our study of push-down automata) which makes the design of efficient parsers a real challenge—how to prune the number of "false leads" to avoid a combinatorial explosion. In fact, the successful reduction in the present example is

$$E + T * a \Leftarrow E + T * F \quad \text{(reverse 6)}$$
$$\Leftarrow E + T \quad \text{(reverse 3)}$$
$$\Leftarrow E \quad \text{(reverse 1)}$$

so that we have a complete derivation, confirming that $a + a * a$ belongs to $L(G)$.

In the preceding example, parsing goes from left to right. The process is, as we have said, a kind of *bottom-up parsing* (working "up the tree" from the string to the start symbol) and is called *right-parsing* because it recaptures a rightmost derivation.

2 Definition. Given a grammar $G = (X, V, P, S)$, a string w in $(V \cup X)^+$ is called a *right sentential form* if there exists a rightmost derivation for w

$$S = w_1 \underset{rm}{\Rightarrow} w_2 \underset{rm}{\Rightarrow} \cdots \underset{rm}{\Rightarrow} w_n = w$$

i.e., each step $w_j \Rightarrow w_{j+1}$ is obtained by replacing the right-most variable in w_j according to some production in P, as indicated by the symbol $\underset{rm}{\Rightarrow}$.

Right parsing tries to reconstruct this derivational sequence in reverse, working from w to S. In particular, then, we can write $w_{n-1} = \alpha A v$ and $w = \alpha \beta v$ where $\alpha \in (V \cup X)^*$, $A \to \beta$ is in P, and $v \in X^*$. Notice that w_{n-1} is itself a right sentential form. We call β a *handle* for w, and say that w can be *reduced* by the reverse application of production $A \to \beta$, to yield $\alpha A v$. To summarize: in order for a substring β of w to be a handle, it must match the right-hand side of some production, its reduction must "lead" back to the start symbol, and it must be the leftmost substring of w that satisfies these properties. Note that the handle and reduction of w will be unique if G is unambiguous, but need not be unique if G is ambiguous.

In the preceding example, it was shown that it is not always clear which substring of a given w is a handle. For example, it took two "false leads" before we discovered that a was the handle of $E + T * a$, and not $E + T$ or T.

The process of right-parsing (also called shift-reduce parsing) can be viewed as taking place on a push-down automaton. We *shift* terminals from the input onto the stack, then *reduce* the handles that we find on top of the stack.

3 Example. In this example we reconsider the grammar of Example 1. Let the stack have symbols $V \cup X$, in this case $\{E, T, F, (,), *, +, a\}$. At any time, the string of symbols on the stack followed by the remaining to-be-read input string will correspond to the current reduced string, hypothesized to be a right sentential form. For example, the reduction $E + a * a \Leftarrow E + T * a$ is to be captured by the following machine transition (there is only one state, so it is not shown):

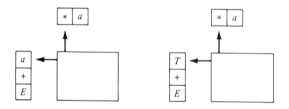

In general, then, each reverse production $\alpha \leftarrow A$ yields a machine transition

$$\langle \lambda, \alpha, A \rangle \quad \text{pop } \alpha, \text{ push } A, \text{ read no input}$$

We also have the transitions

$$\langle x, \alpha, x\alpha \rangle$$

which read a symbol off the input and push it onto the stack. As is clear from Example 1, this machine will in general be nondeterministic. Starting with any empty stack, it will accept strings of $L(G)$ by terminating with the start symbol of G as the sole symbol on its stack.

Our emphasis will now be on ways to avoid the generally nondeterministic "blind search" for possible handles that is present in unrestricted right-parsing. To this end we study $LR(k)$ grammars, a restricted class of context-free grammars which allow us to make a firm decision as to whether a substring of w is a handle by looking ahead to symbols to the right of the substring and by using some *finite-state* information about the symbols to the left of the substring. We thus precede our study of $LR(k)$ grammars with a careful discussion of the concept of *left context* for context-free grammars in general.

4 Definition. Let G be a context-free grammar, with $A \in V$. The *left context of* A,

$$LC(A) = \{w | S \overset{*}{\underset{rm}{\Rightarrow}} wAz, \text{ with } w, z \in (V \cup X)^*\}$$

In words, the left context of a variable A in a grammar G is the set of all strings occurring to the left of A in any sentential form derived in a rightmost derivation from the start symbol. (In general, this is a string of both terminals and nonterminals.)

5 Example. Consider the grammar $G = (X, V, S, P)$ with productions

$$S \rightarrow aAb$$

$$A \rightarrow aAb|ab$$

In this case, by inspection $LC(A)$ is aa^*.

6 Definition. Let G be context-free, and suppose $A \rightarrow w$ is in P. Then we define the left context of a production $A \rightarrow w$ as follows:

$$LC(A \rightarrow w) = LC(A) \cdot \{w\}.$$

7 Example. In Example 2, $LC(A \rightarrow aAb) = aa^*aAb$.

And now we come to the fundamental result of this section: we show that the left contexts of variables (and therefore the left contexts of productions) are regular sets. This result will allow us to use finite state machinery to determine the actions taken by an $LR(k)$ parser. Our technique for building the contexts will use the fixed point methods developed in Section 6.2. Before establishing the general result, we work out an example which will illustrate our method.

8 Example. Consider the matched parenthesis grammar, where for notational convenience we now use square parentheses.

$$S \rightarrow SS|[S]|[]$$

Let us write $\langle S \rangle$ for the left context of S—that is, $\langle S \rangle$ stands for the set of strings over $\{[,],S\}$ that can appear to the left of S whenever it occurs in any rightmost derivation of a string of matched parentheses.

To calculate $\langle S \rangle$, we write an equation that characterizes the terms that contribute to the left context of S. The equation has the structure of an inductive definition. For its basis there is a λ term, which represents the left context of S after length 0 derivations. For the terms representing derivations of length $n + 1$ consider a length n derivation $S \overset{*}{\underset{rm}{\Rightarrow}} \alpha S \beta$. Here α represents an element of the left context of S. If we continue with $S \rightarrow SS$ to get $\alpha SS\beta$—a length $n + 1$ derivation, we see that the first S has α in its left context, while the second has αS in its left context. If we had applied $S \rightarrow [S]$ to get $\alpha[S]\beta$, then this S would have $\alpha[$ in its left context. We thus deduce that

$$\langle S \rangle = \langle S \rangle + \langle S \rangle S + \langle S \rangle [+ \lambda = \langle S \rangle (\lambda + S + [) + \lambda.$$

Now let us look closely at our equation. It represents a left-linear (analogous to a right-linear) grammar, and so if we appeal to Theorem 6.2.6, treating left context markers (symbols in angle brackets) as variables and variables and terminals of the original grammar as the new terminals, we obtain

$$\langle S \rangle = (\lambda + S + [)^* = (S + [)^*$$

That is to say, in any legal rightmost derivation of a string of matched parentheses, the symbol string to the left of any occurrence of S belongs to $(S + [)^*$. (Example: $S \Rightarrow SS \Rightarrow S[S] \Rightarrow S[[S]] \Rightarrow S[[[]]] \Rightarrow [][[[]]].$)

9 Theorem. *Let G be a context-free grammar with no λ-productions and no useless variables (i.e., variables which do not derive terminal strings). Then for any production $A \to w$ of this grammar, both $LC(A)$ and $LC(A \to w)$ are regular sets.*

PROOF. Clearly we need only show that $LC(A)$ is regular. But the left context equation for the variable A in V is given by

$$\langle A \rangle = \langle B_1 \rangle w_1 + \cdots + \langle B_n \rangle w_n + \varepsilon(A)$$

where $\varepsilon(A) = $ **if** $A = S$ **then** λ **else** $\varnothing$, and the $B_i \to w_i A z_i$, $1 \leq i \leq n$ are the productions of G with A on the right-hand side, with a production listed separately for each position of such an A. This is because $\alpha \in \langle B_i \rangle$ iff, in a rightmost derivation $S \overset{*}{\Rightarrow} \alpha B_i \beta \Rightarrow \alpha w_i A z_i \beta$, we also have $\alpha w_i \in \langle A \rangle$. The preceding equations are left-linear by inspection, and so their solution is regular, by the methods of Section 6.2. $\square$

We illustrate the principle of the theorem with this next example.

10 Example. The following grammar is the familiar grammar of Section 4.2, but with $\oplus$ for the Boolean "plus" to prevent confusion with the $+$ used in context equations.

$$S \to R$$
$$S \to [S]$$
$$R \to E = E$$
$$E \to [E \oplus E]$$
$$E \to a$$
$$E \to b$$

We first write the context equations for the grammar:

$$\langle S \rangle = \langle S \rangle [+ \lambda$$

$$\langle R \rangle = \langle S \rangle$$

$$\langle E \rangle = \langle R \rangle + \langle R \rangle (E=)+\langle E \rangle [+ \langle E \rangle ([E \oplus)$$

$$= \langle E \rangle ([+ [E \oplus)+\langle R \rangle (\lambda + E=).$$

Solving—do $\langle S \rangle$ first—we get

$$\langle S \rangle = ([)^*$$

$$\langle R \rangle = ([)^*$$

$$\langle E \rangle = (([)^*(\lambda + E=)) \cdot ([+ [E \oplus)^*$$

Next we tabulate the left context of each production:

$$S \to R \qquad\qquad [^*R$$

$$S \to [S] \qquad\qquad [^*[S]$$

$$R \to E = E \qquad\quad [^*E = E$$

$$E \to [E \oplus E] \qquad ([^* + [^*E=)([\lambda + [E \oplus)^*[E \oplus E]$$

$$E \to a \qquad\qquad\;\; ([^* + [^*E=)([+ [E \oplus)^*a$$

$$E \to b \qquad\qquad\;\; ([^* + [^*E=)([+ [E \oplus)^*b$$

The reader should check that the left contexts of the grammar's productions not only are disjoint but have the property that no string in one context is an initial segment of a string in the context of a different production (see Exercise 1), and this leads us to the following definition.

11 Definition. A grammar $G = (X, V, S, P)$ is $LR(0)$ if for distinct productions $A \to W$, $A' \to W'$ (where A, A' and W, W' need not be distinct so long as the productions are distinct),

$$LC(A \to W) \cap Init(LC(A' \to W')) = \varnothing$$

Thus the grammar of Example 10 is our first interesting example of an $LR(0)$ grammar. When a grammar is $LR(0)$, it is always possible to interpret the current left context of a (backward) rightmost derivation unambiguously and decide whether a parser should shift (that is, read) a next input symbol, or reduce the current stack by using a unique production rule. Before looking at the more general notion of look-ahead, we explore a key implication of Theorem 9.

Let $G = (V, X, S, P)$ be a context-free grammar. There are thus finitely many left contexts $LC(A)$, for A in V, associated with G, and each of these is regular. Let M_A be the finite-state acceptor for $LC(A)$, $M_A = (Q_A, \delta_A, q_{0A}, F_A)$. We can then form the product machine M

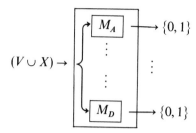

comprising the M_A's operating in parallel on common input

$$M = (Q, \delta, q_0, \beta)$$

where

$Q = \prod_{A \in V} Q_A$, the Cartesian product of the state sets of the M_A's

$q_0 = (q_{0A}|A \in V)$ is the V-tuple of initial states of the M_A's

$\delta((q_A|A \in V), x) = (\delta(q_A, x)|A \in V)$

Instead of giving M a set of accepting states, however, we give it an output map

$$\beta: Q \to \{0, 1\}^V$$

where $\beta((q_A|A \in V)) = (\beta_A(q_A)|a \in V)$ with

$$\beta_A(q_A) = \begin{cases} 1 & \text{if } q_A \in F_A \\ 0 & \text{if not.} \end{cases}$$

Thus $\beta(\delta^*(q_0, w)) = (\beta_A(\delta_A^*(q_{0A}, w))$ and so has A-component equal to 1 just in case $w \in LC(A)$. Set $\gamma_A = \pi_A \cdot \beta$. We have thus proved the following:

12 Theorem. *For each context-free grammar $G = (X, V, S, P)$ there is a finite-state acceptor M with state set Q, initial state q_0, and transition function δ which is equipped with a map $\gamma_A: Q \to \{0, 1\}$ for each A in V, such that*

$$\gamma_A(\delta^*(q_0, w)) = 1 \text{ iff } w \in LC(A).$$

Thus, M receives as input an element of $(V \cup X)^*$—typically an initial segment of some half-parsed string—and returns as output a binary string with 1 in its ith position iff the input string belongs to the left context of the ith variable.

We now return to parsing. In the last section we studied top-down parsing, which we took to be equivalent to extending a leftmost derivation. Thus, in trying to infer the middle step

$$S \overset{*}{\underset{lm}{\Rightarrow}} \alpha A v \Rightarrow \alpha \beta v \overset{*}{\underset{lm}{\Rightarrow}} \alpha w \in X^*$$

we know A and, because we know the string αw that is being parsed, we know that β must be a prefix of the given w. Thus, in $LL(k)$ parsing, we use look-ahead k to choose which of the possible productions $A \to \beta$—where A is fixed and β is known to be a prefix of w—should be applied.

By contrast, in bottom-up parsing we are trying to extend a rightmost derivation *in reverse*. Thus, in trying to infer the middle step

$$wv \overset{*}{\underset{rm}{\Leftarrow}} \alpha\beta v \Leftarrow \alpha A v \overset{*}{\underset{rm}{\Leftarrow}} S$$

(moving from left to right), we do *not* know the value of A, nor do we know how many symbols of $\alpha\beta$ actually constitute the handle β. Consider, then, any other string which agrees with $\alpha\beta v$ for at least k symbols beyond β (or to the end) and which is rightmost derivable from S. We shall require that our finite-state knowledge of α (Theorem 12) together with this look-ahead right of k symbols disambiguate the last production used in the rightmost derivation. What complicates the formal analysis is that this tail-end v may be less than k symbols long, and so we introduce a symbol \$ not in the set $V \cup X$ of nonterminal and terminal symbols of our grammar G. Then we may rewrite the key stage of our rightmost derivation (now expressed from left to right) as

$$S\$^k \overset{*}{\underset{rm}{\Rightarrow}} \alpha A v_1 v_2 \Rightarrow \alpha\beta v_1 v_2$$

where we may now take $|v_1| = k$ since $v_1 v_2 = v\$^k$ is guaranteed to have length at least k.

13 Definition. We say that the context-free grammar G is $LR(k)$ if, whenever we have two rightmost derivations

$$S\$^k \overset{*}{\underset{rm}{\Rightarrow}} \alpha A w_1 w_2 \underset{rm}{\Rightarrow} \alpha\beta w_1 w_2$$

$$S\$^k \overset{*}{\underset{rm}{\Rightarrow}} \gamma A'w' \underset{rm}{\Rightarrow} \gamma\beta'w' = \gamma\beta w_1 w_2'$$

of right sentential forms which agree up to k places beyond β ($|w_1| = k$) and comprise only terminals thereafter, and such that α and γ belong to the same collection of left contexts, then the handles β and β' must be identical:

$$A \rightarrow \beta = A' \rightarrow \beta'.$$

We use C_α to denote the set of variables to whose left context α belongs. Let R_1 be the set of all right-hand sides of productions of G, and let R_2 be the set of all length k strings of $X^* \cdot \$^*$. Then the $LR(k)$ condition says that *given* α we have a map

$$f_\alpha: R_1 \times R_2 \rightarrow V \cup \{\varnothing\}$$

such that $f_\alpha(\beta, w_1) = A$ if there is a rightmost derivation

$$S\$^k \overset{*}{\underset{rm}{\Rightarrow}} \alpha A w_1 w_2 \underset{rm}{\Rightarrow} \alpha\beta w_1 w_2 \text{ for some } w_2$$

while $f_\alpha(\beta, w_1) = \varnothing$ if no such A exists. The $LR(k)$ condition guarantees that the function f_α is single-valued, and that if $C_\alpha = C_\gamma$ then $f_\alpha = f_\gamma$.

The real problem—if we know f_α (but not α itself) and read in a new symbol $x \in (V \cup X)$— can we then compute $f_{\alpha x}$? While f itself is not readily updatable, there is a more complex function N_α which *is* updatable and contains the necessary f_α-information and such that there are only finitely many distinct N_α's for the infinitely many α's. We simply use the state set Q of Theorem 12, and use y_α to denote the state $\delta^*(q_0, \alpha)$. (Notice that if $\delta^*(q_0, \alpha) = \delta^*(q_0, \alpha')$ then y_α and $y_{\alpha'}$ are actually the *same* symbol.) Then y_α determines C_α, and thus f_α. The update rule for y_α is simply that given by Theorem 12: $y_{\alpha x} = \delta(y_\alpha, x)$.

With the problem of defining and updating f_α solved by using y_α as intermediary, we next show how to construct a *deterministic* PDA for accepting $L(G)$ for the $LR(k)$ grammar G.

14 Theorem. *If G is $LR(k)$, then $L(G)$ can be accepted by a deterministic push-down automaton.*

PROOF. If the string w is to be parsed, we provide $w\$^k$ as input to our push-down automaton M. Thus the tape alphabet of M is $X \cup \{\$\}$. Let Y be a finite set of symbols, one for each distinct y_α. (To repeat: we use y_α to denote the symbol corresponding to $\delta^*(q_0, \alpha)$. Thus a given y in Y may be y_α for infinitely many α in $(V \cup X)^*$.) We then stipulate that the stack alphabet of M is $Z = (V \cup X) \times Y$. At any time, then, the ID of the machine will have the following form (Figure 1):

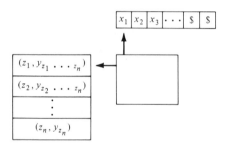

Figure 1

The idea is that, at each step of its operation, the machine will process a string $z_n \ldots z_2 z_1 x_1 x_2 x_3 \ldots \$\$$ which represents a rightmost derivation from $S\$^k$. But in addition to having the letters $z_i \in (V \cup X)$ available on the stack, stack entries also contain left-context symbols $y_{z_i \ldots z_n}$. When the control head reads such a y, it can look up a table which tells it the corresponding f_α. We use this in the reduction process diagrammed by Figure 2. Here we see that the z's on the stack correspond to $\alpha\beta w_1$—i.e., they include the k look-ahead. We want to replace β by A just in case the production $A \to \beta$ is sanctioned by the look-ahead. Clearly, such a replacement must be accompanied by an updating of all the y's corresponding to the z's of A and w_1 on the stack.

In the reduction of Figure 2, then, the control box pops the top of the stack corresponding to the symbols of w_1, β, and the top (last) symbol for α (which tells it y_α); and then pushes the stack symbols for w_1, A and the top element of α, using y_α to compile the correct new y's.

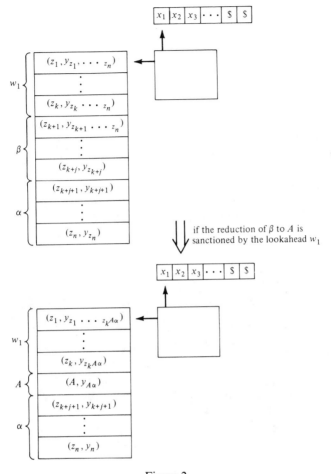

Figure 2

Let N be the length of the longest right-hand side of a production of G. Then for any $j \leq N$, we can decompose the stack from bottom to top as $(z_n, y_{(n)}) \ldots (z_{k+j+1}, y_{(k+j+1)})(z_{k+j}, y_{(k+j)}) \ldots (z_{k+1}, y_{(k+1)})(z_k, y_{(k)}), \ldots, (z_1, y_{(1)})$ so long as $k + j + 1 \leq n$. Here we have used $y_{(r)}$ to abbreviate $y_{z_1 \ldots z_r}$. Let us then set

$$\alpha = z_n \ldots z_{k+j+1}, \ \beta = z_{k+j} \ldots z_{k+1}, \ w_1 = z_k \ldots z_1.$$

Thus α is the left context, and we want to test whether β is indeed the handle for $z_n \ldots z_1 x_1 \ldots \$$, making use of the right context w_1 of k symbols. The crucial point is that M does not have to examine the arbitrarily many symbols that make up α. Instead, it simply reads the top symbol of the α-portion of the stack to find y_α. From y_α it can retrieve the table for f_α and look up $f_\alpha(\beta, w_1)$.

If the result is $\emptyset$, β is not the handle, and M must try another j, or read some more of the input. If $f_\alpha(\beta, w_1) = A$ in V, then β is a handle and may be reduced by applying the production $A \to \beta$ in reverse. Because of our assumption that for any y_α and x we can obtain $y_{\alpha x}$ without knowing α (only y_α), we can provide the necessary $y_{A\alpha}$, $y_{z_k A\alpha}$, $\cdots$ $y_{z_1 \ldots z_k A\alpha}$ for the top $k + 1$ symbols of the stack which sit atop α after the reduction.

Note that the preceding procedure requires the control box to examine at most the top $k + N + 1$ symbols of the stack, and we thus have a legitimate generalized PDA.

We may thus specify the transition function for PDA M as follows:

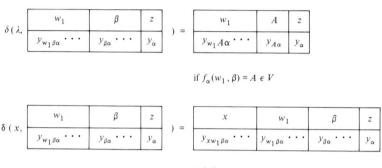

$$\text{if } f_\alpha(w_1, \beta) = A \in V$$

if $|\beta| = N$, and no prefix of β passes the "handle test".

To complete the specification, we specify that the start symbol for the stack is $(\$, y_\lambda)$, and we have the transition

$$\delta(x, \begin{array}{|c|c|c|} \hline w & \$ \\ \hline y_w \cdots & y_\lambda \\ \hline \end{array}) = \begin{array}{|c|c|c|} \hline x & w & \$ \\ \hline y_{xw} & y_w \cdots & y_\lambda \\ \hline \end{array}$$

It is then clear that the specified PDA is deterministic, and it accepts strings of $L(G)\$^k$ by terminating with

$$(\$, y_{\$^k S}) \ldots (\$, y_{\$ S})(S, y_S)(\$, y_\lambda)$$

on its stack.   □

We close this section by noting some surprising consequences of the last theorem.

15 Corollary. *Not every context-free language has an LR(k) grammar.*

PROOF. The preceding result shows that if G is an $LR(k)$ grammar, then $L(G)$ is deterministic. But, as we noted in Section 2.3, *deterministic* languages are closed under complementation—just run the deterministic PDA for the language in the one possible way to decide whether or not a string belongs to the language (i.e., deterministic PDAs are deciders, and not just acceptors)—and thus, form a *proper* subfamily of the context-free languages. $\square$

Extending the observation in the preceding proof, we have the following surprising result:

16 Theorem. *A context-free L has an LR(k)-grammar for some $k \geq 0$ iff it is deterministic iff it has an LR(0)-grammar.*

PROOF OUTLINE. We already know that "$LR(k) \Rightarrow$ deterministic," and (by Exercise 3) we know that "$LR(0) \subset LR(k)$." It only remains to check that "deterministic $\Rightarrow LR(0)$." Suppose that M is a deterministic PDA that accepts L. We then construct the grammar G whose productions mimic the transitions of M in reverse, in the style of Section 3.3. We then note that, because M is deterministic, G is $LR(0)$ since each reduction directly replaces symbols at the top of the stack, and no look-ahead is required at all. $\square$

The price for the transition from an $LR(k)$ grammar to an $LR(0)$ grammar is a great increase in the set of nonterminals, as we pass from V to $V \times Y$ in the style of the proof of Theorem 3.3.4, and this in turn entails a great increase in the number of productions, making such a transformation impractical.

EXERCISES FOR SECTION 7.2

1. Verify that the grammar of Example 10 is $LR(0)$.

2. Is the grammar of Example 1 $LR(0)$?

3. Show that every $LR(0)$ grammar is $LR(k)$ for all $k > 0$.

4. Does every regular language have an $LR(0)$ grammar? What about an $LR(1)$ grammar?

CHAPTER 8

The Formal Description of Natural Languages

8.1 The Aims of Linguistic Theory
8.2 The Generative Power of Natural Languages
8.3 Augmented Transition Networks

As we mentioned in Chapter 1, the modern theory of formal grammars was in large part initiated by Chomsky's analysis of the generative power of natural languages, the languages of everyday human usage. In this chapter and the next we will explore in more detail just what formal power natural languages seem to have, and how this is captured in contemporary linguistic theories. We will see that there is no easy answer to the question, What is the generative power of natural languages? There is still some disagreement, in fact, over whether natural languages are recursive or not.

It is not traditional for computer science textbooks on formal languages to include chapters on natural language. However, we are led to this innovation by two current trends. The first is that artificial intelligence (AI) has now entered the mainstream of computer science, with the design of natural language interfaces being an important area of AI applications. The second is that more and more computer scientists are working with colleagues in linguistics, neuroscience, and psychology in developing the field of cognitive science. For both these reasons, it is increasingly likely that computer scientists will make use of formalizations of natural language as well as programming languages.

We will proceed as follows. After a brief description of the aims of linguistic theory, we review the arguments for denying that natural languages are regular or context-free. Then we will examine the classical theory of transformational grammar (Chomsky 1965) and show that the power allowed by this theory is too great for purposes of describing natural languages. Indeed, we will show that even undecidable sets can be generated by transformational grammars. Then we will demonstrate how a class of automata, *recursive transition networks*, can be modified to *augmented transition networks* in order to parse any natural language. In the next chapter, we examine a recent formulation of context-free grammars which tries to capture the expressiveness of natural languages without transformational rules. Following that, we discuss the contemporary view of transformational grammar as a restricted theory with several constraining interacting subcomponents. Chapter 9 closes with a list of references for material covered in these two chapters.

Among the formalizations we consider are two that have been used by AI researchers in designing natural language interfaces: augmented transition networks and generalized phrase structure grammars. For a view of cognitive models of natural language (parsing, acquisition, and production), see (Arbib et al. 1987).

8.1 The Aims of Linguistic Theory

When we study a programming language we have a clear understanding of the structural complexity of the expressions in that language. This is so because we usually have an explicit statement of the rules of syntax for the language and these rules are frequently accompanied by at least an informal semantics for the legal expressions. By contrast, we have no direct access to the principles and rules of a particular natural language: they are concealed in our cognitive makeup. In fact, natural languages are much more complicated than one might at first glance suspect. In this section we will introduce some basic terminology from linguistics and then show what kind of diagnostics and techniques linguists use to examine the structure of natural language utterances. Then we will articulate what the general and specific goals of a theory of language should be.

Knowledge of language is seen by contemporary linguists as a complex interaction of abilities relating specific linguistic knowledge to other nonlinguistic abilities. Because it is sometimes difficult to study familiar phenomena, however, a certain "psychic distance" from the facts being observed is necessary if we hope to achieve progress in the study of cognitive faculties, and language in particular. It is therefore important to formulate analyses of the observable data that not only account for the facts, but also achieve some sort of explanatory adequacy, revealing principles behind the surface phenomena.

One of the major goals of linguistic theory is to arrive at generalizations about the structure and meaning of sentences, and at least five perspectives contribute to this goal:

(1) Syntax—the structure of a sentence
(2) Semantics—the meaning of a sentence
(3) Pragmatics—the meaning in the context in which it is spoken
(4) Morphology—the structure of a word
(5) Phonology—the structure of sounds

It should also be noted that there are many other concerns in linguistics, such as the acquisition of language by a child, the structure of discourse, and the study of brain mechanisms underlying language. We will, however, only concern ourselves here with the first area, syntax.

Focusing, then, on syntax, it becomes apparent that there are certain structural regularities that show up in the sentences of a language. Some of these patterns are obvious and can be understood immediately. For example, the italicized words in each sentence below appear in the same structural context, and all perform the same *grammatical function* in the sentence, that of *subject*.

(1) *John* left the party early.
(2) *The man with the coat* left the party early.
(3) *Every guest* left the party early.
(4) *He* left the party early.

Similarly, in the following sentences, these words appear in a new configuration in the sentence to perform the grammatical function of *object*.

(1) Mary visited *John*.
(2) Mary visited *the man with the coat*.
(3) Mary visited *every guest*.
(4) Mary visited *him*.

Each one of these italicized strings can be replaced by any of the others and the sentence will remain grammatical. This is part of the *distributional analysis* that linguists perform to establish constituency of words or phases. The unit that is important in this case is called a *noun phrase*. Similar distributional analyses can be applied directly to establish the constituency of *prepositional phrases*, *adjectival phrases*, and *verb phrases*, as illustrated in the following:

PREPOSITIONAL PHRASE

(1) The man *with the coat* walked in.
(2) The book *on the shelf* is mine.
(3) John put the book *on the shelf*.

ADJECTIVAL PHRASE

(1) The *young and happy* couple just got married.
(2) My children are *young and happy*.

VERB PHRASE

(1) Bill *ate the cake* and Mary *ate the pie*.
(2) Mary *likes to go swimming* and Bill *does too*.
(3) John made Mary *pack her bags*.

While this seems like a straightforward procedure for determining constituency, other structural differences and similarities in a sentence may not be as apparent, and the process by which we extract this information may be more elaborate. Consider, for example, the following two sentences which are identical except for the main (or what we will call the *matrix*) verb.

(1) I *expected* John to leave.
(2) I *persuaded* John to leave.

What we know about these sentences is that there is a clear grammatical relation between *John* and *to leave* in each sentence, something we will term a *subject-predicate* relation. It also seems clear that in (2) *John* is also functioning as the object of *persuade*. while in (1) *John* performs no such role. Informally, the difference can be seen by paraphrasing the two sentences: in (1), what is being expected is the entire proposition, "John's leaving"; in (2), however, John and John alone is being persuaded.

There are many syntactic tests (also known as *diagnostics*) which indicate that these two sentences require different structural analyses. One such test involves what is called the *selectional properties* of a verb. Notice that the verb *persuade* requires that the noun phrase appearing as the subject of the subordinate verb must be animate. That is, one cannot persuade a book to do something because it is not animate. This type of dependency between a verb and its object is called a *selectional restriction*. The verb *expect* has no such restriction on the subject, however. Hence, we get the following contrasts, where "*" means ungrammatical.

(1) *I persuaded the book to fall from the shelf.
(2) I expected the book to fall from the shelf.
(3) *I persuaded it to rain.
(4) I expected it to rain.
(5) *I persuaded headway to be made on the project.
(6) I expected headway to be made on the project.

Another way in which these verbs differ is illustrated by the following contrast.

(1) *It is persuaded that John will come.
(2) It is expected that John will come.

That is, *except* permits this special passive form while *persuade* does not. Another set of facts that seems to indicate differences between these two verbs comes from examining the meanings of the following sentences.

(1) I persuaded a specialist to examine John.
(2) I expected a specialist to examine John.
(3) I persuaded John to be examined by a specialist.
(4) I expected John to be examined by a specialist.

Sentences (2) and (4) show no difference in meaning, whereas there is an important difference between (1) and (3): two different people are being persuaded in (1) and (3), yet in (2) and (4) the same event is expected. This could be explained if different structures are assigned to the sentences based on the selection of the verb.

Thus although these two sentences show superficial similarities in structure, there is strong reason to believe that there are very different structures associated with each. Essentially, the difference is in what is structurally considered the object of the verb. For *expect*, the object is an entire sentence, while *persuade* seems to have two objects (also called *complements*), a noun phrase and an embedded sentence (with a subject and a predicate).

The tree structures we first introduced in Section 1.3 are the key to the analysis of natural language syntax. Consider, for example, the tree shown next:

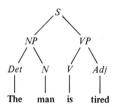

where we use the following notation: S: sentence, NP: noun phrase, Det: determiner, Adj: adjective, N: noun, and VP: verb phrase. This tree allows us to think of the sentence as built up by nesting phrases of increasing complexity: "man" is part of "the man" is part of "the man is tired." A *phrase marker* is a string notation which makes explicit this phrase structure. Linguists often display such trees using this notation:

$$[_S[_{NP}[_{Det}\text{the}][_N\text{man}]][_{VP}[_V\text{is}][_{Adj}\text{tired}]]]$$

The inductive definition of a phrase marker is very simple. If t is a tree, and $\{t\}$ is the phrase marker for that tree, then

(1) the phrase marker for a simple word is the word itself
(2) the phrase marker for

is $[_A\{t\}]$
(3) the phrase marker for

is $[_B\{t_1\}\ldots\{t_n\}]$

Let us return briefly to the structural distinction between *expect* and *persuade* in terms of phrase markers. As noted, *persuade* has two objects whereas *expect* has only one.

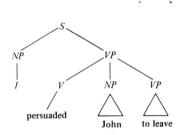

$[_S[_{NP}I][_{VP}[_V persuaded][_{NP}John][_{VP}to leave]]]$

(a)

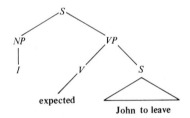

$[_S[_{NP}I][_{VP}[_V expected][_S[_{NP}John][_{VP}to leave]]]]$

(b)

Actually the preceding trees are only the crude outline of a full syntactic analysis. The reader will probably find it confusing that in the latter analysis we classify "John to leave" as a sentence, *S*. For a full discussion of this see (van Riemsdijk and Williams 1986).

The tests described for determining the constituency of phrases and the behavior of certain words provide us with diagnostics for the proper structural analysis of sentences, which we can use for testing our theories of underlying linguistic principles empirically.

In addition to studying the ordering and structural dependencies within a single sentence, as was described above, linguistic theory attempts to explain the relatedness of different sentences that "have the same meaning." That is, the theory should capture the fact that speakers of a language have intuitions about one sentence meaning the same as another, even though they are not structurally identical. Perhaps the best introductory example of this is the relation between the active and passive forms in English.

(1) John washed the car.
(2) The car was washed by John.

Except for certain shifts in emphasis, most speakers of English agree that these two sentences *mean* the same thing: namely, that there was a *washing*, and that John did it, and that the car was the object of this act. This notion of *sameness* of meaning has nothing to do with the word order of the two sentences, however, but rather with the *argument structure* involved in both sentences. That is, *John* is the agent of the washing in both sentences just as *the car* is the patient, or the object acted upon.

Another example of this relatedness between sentence forms is seen in the following pair of sentences, which unquestionably have identical meanings.

(1) John gave the book to Mary.
(2) John gave Mary the book.

Once again, the sameness of meaning comes from the fact that each noun phrase performs the same role in both sentences.

In his early work, Chomsky proposed that this relatedness between sentences be captured in linguistic theory by deriving the two phrase markers, called the *surface structures (S-structures)*, for a given pair of sentences from the same underlying form, called the *deep structure (D-structure)* of each sentence. The surface difference arises from the application of structure-changing operations, called *transformational rules*, to which we will return shortly.

A word of caution: The theory of transformational grammar has evolved greatly in the last 20 years (cf. Section 8.3). Thus, the account of active and passive suggested here is historically important but is no longer part of the current theory (see (van Riemsdijk and Williams 1986) for a very comprehensive and clear treatment of how the theory has changed).

In order to understand the exact nature of transformational rules, it will be important to examine how the different components of a linguistic theory interact to derive a surface structure. In the early formulation of transformational theory (Chomsky 1965), the grammar for a natural language was seen as having the following components:

(1) A set of phrase structure rules
(2) A lexicon (the dictionary for the language)
(3) The transformational rules
(4) Rules of phonology

All phrase structure rules are considered to be context-free rules of the form given in the following fragment.

$S \rightarrow NP\ Aux\ VP$
$NP \rightarrow Det\ N|Det\ AP\ N|Proper\ N$
$AP \rightarrow Adj\ AP|Adj$
$VP \rightarrow V\ NP|V|V\ PP$
$PP \rightarrow Prep\ NP$

where the symbols are defined as follows:

Adj—Adjective
Aux—Auxiliary
AP—Adjectival phrase
Det—Determiner
N—Noun
PN—Proper noun
NP—Noun phrase
Prep—Preposition
PP—Prepositional phrase
S—Sentence
V—Verb
VP—Verb phrase

In Chapter 1 we viewed the lexicon as simply a listing of *preterminal rewrite rules*, such as the following:

$V \rightarrow sees|eats|runs|walks\ldots$
$Det \rightarrow the|its|this|a|every\ldots$
$Prep \rightarrow on|in|for\ldots$
$N \rightarrow fish|woman|man|park|dog\ldots$
$Adj \rightarrow big|red|small|proud\ldots$

However, generative grammar posits a much richer structure for lexical items of a language. The lexicon contains the following type of information:

(1) *Categorization*
(2) *Subcategorization*
(3) *Selectional restrictions*
(4) *Argument structure*
(5) *Lexical semantics*
(6) *Phonetic representation*

We illustrate (1), (2), and (4) with the lexical entry for the verb *hit*.

$$hit: V, \langle NP \rangle, (AGENT, THEME, INSTR)$$

The types of information represented here include categorization, subcategorization, and argument structure. The categorization of *hit* is simply a specification of the grammatical type, *V*. The subcategorization for *hit*, $\langle NP \rangle$, is a way of specifying that *hit* is a transitive verb, taking an *NP* object. Thus, within the category of verb, for example, there are several subcategories of verbs, some that are intransitive like *walk* (taking no object, $\langle \rangle$), others are transitive, like *hit* (taking one object, $\langle NP \rangle$), and there are still others taking prepositional phrases, etc. The argument structure for *hit* is simply a list of the *thematic roles* associated with this verb. These include roles such as, *AGENT, GOAL, SOURCE, THEME, LOCATION, INSTR* (intrumental), etc. For example, in the sentence

Mary hit Bill with a baseball bat.

Mary acts as the *AGENT* of the event, *Bill* as the *THEME*, whereas *a baseball bat* is the instrument of the hitting, the *INSTR*. Whenever *hit* is used, other elements of the sentence (perhaps unspecified elements) must play these roles.

The phrase structure rules and the lexicon together make up what is called the *base component*, and it is from this that the deep structure of a sentence is formed. For example, the underlying deep structure for both active and passive forms for a sentence will be the same, as illustrated by the analysis of the following sentences:

(1) This man loves Mary.
(2) Mary is loved by this man.

SURFACE STRUCTURES

(1′) $[_S[_{NP}$this man$][_{Aux}][_{VP}[_V$loves$][_{NP}$Mary$]]]$

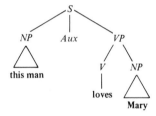

(2′) $[_S[_{NP}$Mary$][_{Aux}$is$][_{VP}[_V$loved$][_{PP}$by this man$]]]$

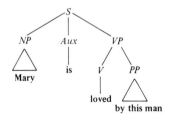

Except for a few details that need not concern us here, these are the surface structures, or *S*-structures, for the sentences in (1) and (2). Note the notation $[_{Aux}]$ in (1′) to show that the auxiliary verb is absent. The corresponding deep structure, or D-structures, for these sentences is the same.

DEEP STRUCTURE

(3) $[_S[_{NP}$this man$][_{Aux}][_{VP}[_V$love$][_{NP}$Mary$]]]$

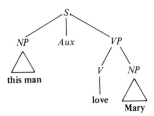

It is at D-structure that the thematic (or semantic) roles are assigned to the different noun phrases. In the D-structure in (3), the object of the verb *love*, *Mary*, is interpreted as the "lovee," or *THEME* in our terms, while the *NP* appearing in subject position at D-structure is interpreted as the "lover," or *AGENT* of the loving. These roles will stay with their associated *NP*s and cannot be reassigned.

Now let us informally introduce the notion of a transformational rule. As was pointed out earlier in the chapter, transformations are useful for establishing generalizations in language, and in this particular case, one rule will capture the relatedness between the active and passive forms of a sentence. This rule, the *passive transformation*, can be roughly stated as follows: In a context, *NP V NP X*, transpose the two *NP*s and add the relevant form of the verb *be*, and change the verb to its past participle. The standard notation for this *structural description* and the accompanying change is given next. X simply indicates that anything can follow the *NP*, and the rule will still apply.

Passive Transformation

SD:	NP	V	NP	X	
	1	2	3	4	$\Rightarrow$
SC:	3	be + 2[pp]	4	by + 1	

The SD, or *structural description*, is the context that must be satisfied for the rule to apply, in this case an *NP* followed by a *V* followed by *NP*. SC stands for the *structural change*. It gives the new structure. The numbers then refer to the corresponding partition of the sentence: 1 denotes the first *NP*, 2 the *V*, 3 the second *NP*, and 4 the *X*. With this, the structural change can be understood as saying that the transformed structure comprises the second *NP* (3), followed by the past participle (pp) of the verb (2), preceded by the relevant form of the verb *be*, followed by *X* (4), then *by*, and then the second *NP* (1). Note that this does not tell us how the relevant form of *be* is chosen.

Let's see how this rule applies to the particular sentence we encountered earlier. In the D-structure,

$$[_S[_{NP}\text{this man}][_{Aux}][_{VP}[_V\text{love}][_{NP}\text{Mary}]]]$$

we partition according to the passive rule, getting

SD:	this man	love	Mary	X	
	1	2	3	4	⇒
SC:	3	be + 2[pp]	4	by + 1	
	Mary	is loved	λ	by this man	

That is, we take 1 = 'this man', 2 = 'love', 3 = 'Mary', while 4 is empty.

From a mathematical point of view, this description is unsatisfactory, despite its acceptance by linguists. This is so because it only describes the *rearrangement of a string* of symbols, rather than making explicit the *restructuring of a tree*. However, we have chosen not to present the proper theory of tree transformations in the present exposition.

Without considering the finer points of agreement and tense, the passive transformation generates the passive form of the sentence, with exactly the same meaning as the active form, which is obtained from the deep structure more directly. The interesting thing about this rule is that it makes no reference to the conceptual or thematic properties of the verb. Its formulation is purely syntactic. What this means is that any D-structure meeting the structural description for this rule should be passivizable. Let us consider an unlikely candidate for the application of this rule, a sentence containing an *idiom*.

(1) The scientists made headway on the project.

Now according to our passive rule, there should be a legitimate passive form for this sentence. The partitioning in the structural description and change is illustrated next.

SD:	the scientists	made	headway	on the project	
	1	2	3	4	⇒
SC:	3	be + 2[pp]	4	by + 1	
	headway	was made	on the project	by the scientists	

The sentence resulting from the application of this rule is in fact grammatical.

> Headway was made on the project by the scientists.

It seems that the passive in English is a very syntax-directed rule and has very little relation to conceptual notions, and this is substantiated by other such examples, where the syntax seems to run "blindly."

(1) John took advantage of the situation.
(2) Advantage was taken of the situation by John.
(3) The situation was taken advantage of by John.

In this case, the partitioning in the passive rule has the option of treating *take* as a normal verb form, in which case we get (2), or the rule can partition *take advantage of* as the verb, and passivize the *NP the situation*.

Given what we have discussed so far, it is difficult to recognize immediately how theoretical results in linguistics can be substantiated in the language we speak every day. Indeed, it sometimes seems as though the product of linguistic theory is more academic than real. There are, however, some striking corroborations of the theory in our daily language.

First some terminology should be made clear. The notion *grammar* that we are using here means something quite different from what we are used to from grade school English and composition classes. There we were taught how we should use language, indicating what is "bad" English and what is standard. This is called *prescriptive grammar*, since it is a prescription on how to use the language. *Universal grammar* is a special term adopted by Chomsky to refer to the innate ability each human has to speak any language. Just as we are genetically endowed for walking on two feet, or seeing with binocular, stereoscopic, vision, we are genetically equipped for the use, production, and processing of language. Note, however, that some cognitive scientists find Chomsky's use of the term *universal grammar* to refer to the genetic language endowment to be highly controversial, for it suggests that what is innate is more like a transformational grammar than, say, a set of learning mechanisms which can infer a grammar from the child's experience with language as a medium of communication. For an example of the latter viewpoint, see (Hill and Arbib 1984). In any case, we may agree on the search for universal grammar as a chore for determining a set of formal grammars particularly well adapted to capturing many of the structural properties of natural languages.

One of the most striking illustrations of results from theoretical linguistics can be seen in the following explanation of the way speakers of English use the common verb *want* in everyday speech. Consider these sentences.

(1) I *want* to buy the beer for the party.
(2) They *want* to leave New York.

Now it is well known that speakers tend to run words together in certain contexts; thus we also hear the following:

(1′) I *wanna* buy the beer for the party.
(2′) They *wanna* leave New York.

This phenomenon is called *contraction*, where *want* and *to* become one word phonetically. This occurs with other verbs, as well: for example,

$$going \; to \rightarrow gonna$$

$$used \; to \rightarrow useta$$

Now consider two sentences that are almost identical to the previous sentences but have an additional noun phrase next to the verb *want*.

(3) I *want* Bill to buy the beer for the party.
(4) They *want* me to leave New York.

There is no possibility of contracting the *want* and *to* in these cases, however:

(3′) *I *wanna* Bill buy the beer for the party.

(4′) *They *wanna* me leave New York.

It seems that *want* and *to* can contract only if they are contiguous (that is, next to each other in the sentence). If something intervenes then the rule cannot take place. So much seems obvious. Let us state this generalization rather schematically as follows:

$$X \text{ want } Y \text{ to } Z \Rightarrow X \text{ wanna } Z$$

only if $Y = \lambda$. That is, contraction applies when nothing comes between the two elements. The real power of linguistics comes into play when we probe how general its application may be. In the following sentences the rule applies without any problem.

(1) What do you *want* to buy for the party?

(1′) What do you *wanna* buy for the party?

(2) What city do they *want* to leave?

(2′) What city do they *wanna* leave?

Something strange happens, however, when we try to contract in the next two sentences.

(3) Who do you *want* to buy the beer for the party?

(3′) *Who do you *wanna* buy the beer for the party?

(4) Who do they *want* to leave New York?

(4′) *Who do they *wanna* leave New York?

This is puzzling since the description of the contraction rule fits; namely, *want* and *to* are contiguous in both sentences. Why then can't contraction take place here? An explanation of this puzzle is found in a richer description of the sentence's structure. Let us see how.

Earlier we introduced the notion of a transformation as a rule taking a phrase structure, partitioning it, and changing it to another phrase structure. This was seen as a way of capturing significant generalizations in the language, such as the relation between active and passive forms of a sentence. Another type of generalization that can be captured by transformations involves sentences like the preceding questions. These questions involving *who* and *what* are called *wh*-questions and exhibit some interesting properties. To see what these properties are, let us compare two sets of sentences, the following indicatives, with some *wh*-question versions derived from them.

Type I

(1a) John drove his car.

(2a) John thinks Mary drove his car.

(3a) John thinks Mary wants Bill to drive her car.

TYPE II

(1b) What did John drive?
(1c) Who drove his car?
(2b) Who does John think drove his car?
(2c) What does John think Mary drove?
(3b) Who does John think Mary wants to drive her car?
(3c) What does John think Mary wants Bill to drive?

How could the sentences in these two classes be generated? It seems that for each sentence of type I, there will have to be a phrase structure rule that generates almost exactly the same structure, the corresponding sentences of Type II, but for the *wh*-element that is in front. This raises two problems. First, there is a great deal of redundancy in having two entirely different sets of phrase structure rules which differ so minimally. Second, we are not capturing in any way the generalization that these pairs are related. Notice that in sentence (1a), the *NP his car* is interpreted as the thing being driven, the *THEME*. In sentence (1b), notice that there is no overt *NP* in the object position, but that we construe the word *what* as asking about the thing being driven, *i.e.*, the *THEME*. Thus, these two sentences are semantically related, but there is nothing in the grammar developed so far to tell us that this connection exists.

One way of capturing this generalization and making it unnecessary to posit two sets of rules is to relate an indicative and a *wh*-question interrogative by a transformational rule, called *wh-movement*. Informally, the rule takes a phrase structure with a *wh*-word in it, and moves that word to a designated spot in the structure. For example, take the D-structure of a sentence like (1b).

(1b′) $[_S[_{NP}\text{John}][_{Aux}][_{VP}\text{drive}[_{NP}\text{what}]]]$

Glossing over a few details, the rule *wh*-movement will take any *wh*-word it finds and move it to the front according to the following change.

	wh-movement			
SD:	X	wh	Y	
	1	2	3	⇒
SC:	2	*do* + 1	*e*	3

The symbol *e* marks an interesting concept developed within linguistic theory. The idea is that *e* is a "trace" left behind from the movement. The trace is an *empty category*, in the sense that we can't hear it spoken, but it has the same categorical status as the element that moved, e.g., *NP*. It encodes on the S-structure our mental knowledge of the exact "role" to which the *wh*-word refers.

Applying this rule to the D-structure in (1b') we get the following partition.

SD:	John drive	what	λ		
	1	2	3		$\Rightarrow$
SC:	2	$does + 1$	e	3	

(4) What$_i$ does John drive e_i?

Here we use the index i to indicate that the *wh*-word and the trace refer to the same thing. The relation between *what$_i$* and e_i is often referred to as a *long-distance dependency*, because the *wh*-phrase may reference a trace that is located at a distant point in the sentence.

The surface structure that results, (4), will have a semantic interpretation that is similar to the indicative, except that there is a *wh*-element in it. In fact, the trace acts as a place holder for the moved element, so that the word *what* can receive a semantic role, namely *THEME* in this case. Comparing the two sentences, we have the following semantic representations for (1a) and (1b), respectively.

(1) DRIVE(John, John's car)
(2) ?(For which x) DRIVE(John, x)

Let us now return to the contraction facts from our earlier discussion. Recall the puzzling fact that we did not get contraction in the following cases.

(1) *Who do you *wanna* buy the beer for the party?
(2) *Who do they *wanna* leave New York?

Our rule of contraction applies to all contiguous (*want, to*) pairs, and yet these are bad sentences. Either our rule was misformulated, or these two elements are not contiguous. In fact, as we have just seen, they are *not* contiguous, since there is a *NP*-trace in between them. Notice that (1) is derived from the D-structure in (1').

(1') You want *who* to buy the beer for the party

When *wh*-movement applies to this structure to give us (1), the movement "leaves a trace." That is, there is still an empty NP in this position. This is made clearer in (2), the resulting surface structure for (1'), where e_i is the empty NP category.

(2) *who$_i$* do you want e_i to buy the beer for the party

Since the contraction rule applies only on string adjacent items, the rule cannot apply.

$$X \text{ want } Y \text{ to } Z \Rightarrow X \text{ wanna } Z$$

only if $Y = \lambda$. But in this case, $Y \neq \lambda$: it contains the *NP* empty category left by the movement transformation.

Hence we see a puzzling fact explained by an independently motivated aspect of our grammatical theory, the notion of empty category. There are many such puzzles in our everyday language that go unnoticed by the speakers of that language and yet which must be explained by the rules and principles of any satisfactory grammar for a natural language.

The final point we will address in this section concerns the *cyclic* nature of transformational rules. Earlier we introduced transformational rules without explicit consideration of ordering or the scope of the rule involved. For example, the passive rule was given as:

Passive Transformation

SD:	NP	V		NP	X	
	1	2		3	4	$\Rightarrow$
SC:	3	be + 2[pp]		4	by + 1	

Now consider a sentence such as the following:

(1) John is believed to be wanted by the police, by everyone in this room.

What exactly is the deep structure for this sentence? The sentence is a passive form, where it appears that the *by*-phrase *by everyone in this room* is the deep subject of the sentence. Thus, this passive is derived from the structure in (2).

(2) Everyone in this room believes John to be wanted by the police.

Notice, however, that this sentence itself has a passive form embedded within the complement of *believe* (i.e. the embedded sentence). So, in fact, sentence (2) is actually derived from the structure given in (3).

(3) Everyone in this room believes the police to want John.

This is the original deep structure for the sentence in (1), shown bracketed in (4):

(4) $[_S[_{NP}$everyone...$][_{VP}$believe$[_S[_{NP}$the police$]$ want John$]]]$

The important thing to notice here is that, in order to arrive at the sentence in (1) by the application of passive, the noun phrase *John* must be in the position immediately to the right of the verb *believe* prior to the passive rule. But notice that in (4) this is not the case. In fact, the only way that the *NP John* can arrive in that position is by passivization on the lower sentence *the police want John*. In this way, the output of one application of passive is the structural description for another application of the same rule. How do we know where to apply the rule first, however?

A crucial part of transformational grammar has been the notion that transformations apply to the lowest (or most deeply embedded) sentence (or *cycle*) first and work their way up to the top cycle, the matrix sentence. In this example, the passive rule will apply to the lower sentence first, with the following effects.

SD:	the police	want	John	X	
	1	2	3	4	⇒
SC:	3	be + 2[pp]	4	by + 1	
	John	is wanted	λ	by the police	

Now we move to the next cycle, or the higher sentence, and apply the Passive Rule again, in the following fashion.

SD:	everyone...	believe	John	X	
	1	2	3	4	⇒
SC:	3	be + 2[pp]	4	by + 1	
	John	is believed	by the police	by everyone...	

So a rule may apply at any or every cycle, if its structural description is met.

Another rule that applies cyclically is *wh*-movement. In fact, it is with this rule that we see the strongest effects of the cycle. Recall that our rule moved a *wh*-phrase to the front of a structural description while leaving a trace of the phrase being moved.

Wh-Movement

SD:	X	wh	Y		
	1	2	3		⇒
SC:	2	do + 1	e	3	

In sentences (1) and (2), the complement phrase has a *wh*-phrase inside it.

(1) John wonders [[who] [Bill wants to marry]]
(2) Mary knows [[who] [Bill thinks [is coming for dinner]]]

Now let us see how a sentence such as (3) is derived. The deep structure is shown as (4).

(3) [*Who_i*] [does Mary believe [John likes e_i]]
(4) [Mary believe [[John likes *who*]]

As noted earlier, we can specify transformations in string notation as in (5).

(5) John likes *who* X ⇒ *who* does John like *e* X

We assume that *wh*-movement applies cyclically, starting with the lower cycle. Also, assuming that we have a rule that deletes *do* if it is not in the highest cycle, this gives the intermediate structure in (6).

(6) [Mary believe [[*who*] [John likes *e*]]]

The rule then applies again on the higher cycle, to give the surface structure (7).

(7) [*Who_i*] [does Mary believe[[e_i] [[John likes e_i]]]

The significant observation captured by the *wh*-rule is that what appears at first to be a sort of "unbounded" movement of the *wh*-element is in fact a local, hopping rule, that keeps applying iteratively until it "lands" in place. We will compare this to a different approach in the next chapter, when we consider alternative models to this theory.

EXERCISES FOR SECTION 8.1

1. Examine the sentences in (1) and (2).
   (1) a. John ate the sandwich.
       b. The woman saw the robber.
       c. Bill visited his grandmother.
   (2) a. It was the sandwich that John ate.
       b. It was the robber that the woman saw.
       c. It was his grandmother that Bill visited.
   Describe the relationship between the sentences in (1) and (2). Write a transformational rule that operates on a structure like that in (1) to give a structure like that in (2).

8.2 The Generative Power of Natural Languages

In the previous section we discussed the linguistic issues which motivated transformational grammar. In this section we explore two issues which tie into formal language theory as we have examined it in the previous chapters. First we examine the generative power of natural languages, and then we consider the inherent generative capacity of classical transformational grammar as a formalism for language competence. We start with the question, Where in the following Chomsky hierarchy do natural languages seem to fit? As we recall from Section 2.1, the hierarchy takes the following form:

$$\mathscr{L}_3 \subsetneqq \mathscr{L}_2 \subsetneqq \mathscr{L}_1 \subsetneqq \mathscr{L}_0.$$

where $\mathscr{L}_3$ is the set of regular languages, $\mathscr{L}_2$ the context-free languages, $\mathscr{L}_1$ the context-sensitive languages, and $\mathscr{L}_0$ the class of all phrase structure languages.

In an early paper Chomsky showed (Chomsky 1959) that natural languages could not be described as regular languages because natural language grammars must exhibit self-embedding. To get the argument going, we assume that English has a context-free grammar, although this is a matter of debate. However, if English were regular, a fortiori it would be context-free, so the point is well taken for our present proof by contradiction.

1 Definition. A CF grammar is self-embedding if there exists $A \in V$ such that

$$A \stackrel{*}{\Rightarrow} \alpha A \beta$$

for some $\alpha, \beta \in (V \cup X)^+$.

2 Theorem. *A CF language L is regular iff it possesses at least one grammar which is not self-embedding.*

PROOF. See Exercise 3.2.11.   □

Note that a regular language may certainly have a self-embedding grammar. For let G_1 be any grammar for X^*, and let G_2 be any self-embedding grammar at all (e.g. $S \rightarrow ab$, $S \rightarrow aSb$), then the grammar G whose productions are the union of those of G_1 and G_2 is clearly self-embedding and is also a grammar for the regular set X^*

Now it is very easy to find examples of self-embedding in natural languages. For example, in English, there are sentences of the form

(1) John believes that Mary wants Bill with all his heart.
(2) John believes that Mary want Bill to leave with all his heart.
(3) John believes that Mary wants Bill to tell Sam to leave with all his heart.

To further the argument, we must note an important methodological assumption in developing grammars in the style given here. It might be argued that everything important about the grammar of English can be restricted to statements about well-formed English sentences containing at most 500 words. These constitute a finite (though enormous) set and are thus a regular set. However, most linguists argue that this length restriction could not be a part of the grammar proper, but merely a restriction inherent in the *use* of the grammar. For example, English grammar needs rules for how to form the conjunction *A and B* from the sentences *A* and *B*, but it would not seem a proper part of the grammar to add the restriction "and these rules may only be applied when *A* and *B* contain at most 499 words between them." In the same way linguists seek rules for generating sentences like (1) to (3) without any restriction on the depth of nesting involved. This forces *any* grammar to allow derivations of the form

$$A \overset{*}{\Rightarrow} \alpha A \beta$$

so that English is self-embedding. Thus, even if the grammar generating these sentences were context-free, we can conclude that English is not regular.

There is a second argument that natural languages are not regular. This conclusion can be reached by simply considering the case of substitution classes in a language. Following Myhill (1957)—see also (Rabin and Scott 1959)—we introduce:

3 Definition. We say that two strings w_1 and w_2 are *Myhill equivalent* with respect to the language L, $w_1 \equiv_L w_2$, if for all strings u, v of X^* we have that

(1) uw_1v is in $L \Leftrightarrow uw_2v$ is in L.

4 Proposition. (a) *If w_1 is in L and $w_1 \equiv_L w_2$, then w_2 is in L.*
          (b) *If $w_1 \equiv_L w_2$ and $x \in X$, then $w_1 x \equiv_L w_2 x$.*

PROOF. (a) Take $u = v = \lambda$ in (1).

(b) We have for any u, v in X^* that

$$u(w_1 x)v \in L \Leftrightarrow uw_1(xv) \in L$$

$$\Leftrightarrow uw_2(xv) \in L \text{ since } w_1 \equiv_L w_2$$

$$\Leftrightarrow u(w_2 x)v \in L$$

Thus, $w_1 x \equiv_L w_2 x$.   □

5 Theorem. *A language L is regular if and only if the number of Myhill equivalence classes for L is finite.*

PROOF. We give the proof in "hard direction," leaving the converse to the reader. Assume that L has a finite set Q of equivalence classes. We will use these classes as the states of an FSA. By Proposition 4, the following definitions for $\delta: Q \times X \to Q$ and $F \subset Q$ are well defined; i.e., they do not depend on the choice of representative w from the equivalence class $[w]$ in Q.

$$\delta([w], x) = [wx]$$

$$[w] \in F \Leftrightarrow w \in L$$

If we now let $q_0 = [\lambda]$, the Myhill equivalence class of the empty string, we have that $M = (Q, q_0, \delta, F)$ accepts L: For clearly

$$\delta^*(q_0, w) = [w]$$

and thus $w \in T(M)$ iff $[w] \in F$ iff $w \in L$.   □

An example of when the preceding finiteness property is violated can be seen with the language $\{a^n b^n | n \geq 1\}$, for clearly b^n is in a different equivalence class for each choice of n. A dependency of this kind can be found in natural language in the following configurations:

(1) The dog died.

$$[_S NP\ VP]$$

(2) The boy that the dog bit died.

$$[_S[_{NP}NP\ [_S NP\ VP\ ]]\ VP]$$

(3) The boy that the dog that the horse kicked bit died....

$$[_S[_{NP}NP\ [_S\ [_{NP}NP\ [_S NP\ VP\ ]]]\ VP]\ VP]$$

where the a's are the NP's and the b's are V's, thus suggesting that English must have infinitely many Myhill equivalence classes and so is not regular.

We now demonstrate that, without adequate restrictions on the types of transformations that can be performed on a string, undecidable sets can be generated by a transformational grammar. Thus, not only do transformational grammars generate the class of what we would call natural languages, but these grammars also generate some very unnatural languages indeed. We will outline how this can be proved, but will not go into the details of the proof (cf. (Kimball 1967) for another proof, and (Salomaa 1971) for an even stronger result).

In order to generate certain important constructions a new type of rule was introduced, a *transformational* rule, which alters the structure generated by phrase structure rules, by moving, adding or deleting material in the string. We mentioned some specific rules for English in the last section, but in principle there are no restrictions on the form of a transformational rule. It is this lack of implicit restrictions that allows a generative grammar to be so powerful in terms of its generative capacity.

In general, a transformational grammar TG has three parts: a phrase structure grammar G, called the *base*, a set of transformations T, and a set of restrictions on these transformations, R. Briefly, the set of derivation trees generated with G provides the set of deep structures of TG. The restrictions of R specify that some transformations in T are obligatory; i.e., they must be applied wherever possible; they also specify the order in which transformations are to be applied. The set of surface structures of TG is the set of trees which may be obtained from deep structures by successively applying transformations from T according to the rules of R so that, in particular, no more transformations need be applied. $L(TG)$ is then the set of strings we may read off these surface structures.

In the next result we consider grammars with *context-free* bases; at the end of the proof we outline the necessary restrictions. We will facilitate our understanding of this result by proving two related lemmas.

6 Lemma. *A transformational grammar can perform arbitrary homomorphisms, and in particular, λ-homomorphisms.*

PROOF DEMONSTRATION. Recall that in Section 2.2 we defined a substitution map $g: X^* \rightarrow 2^{Y^*}$ as a map where $g(\lambda) = \lambda$ and for each $n \geq 1$, $g(a_1 \ldots a_n) = g(a_1)g(a_2)\ldots g(a_n)$. If $g(a)$ contains only one element of Y^* for each $a \in X$, then g is called a *homomorphism*. We can write transformations that perform such mappings over the strings of a language. As an example, imagine a transformation that changes every occurrence of "that" to "my own," and "dog" to "white cat." That is,

$$g(that) = \text{my own}$$

$$g(dog) = \text{white cat.}$$

Then, for some grammar G generating the following sentence,

That dog likes that food.

the resulting string will be as follows:

My own white cat likes my own food.

When we allow λ-homomorphisms, that is $g(a) = \lambda$, we can do arbitrary deletion on the strings generated by the grammar. For example, if we were to add

$$g(that) = \lambda$$

to the original grammar, then the first sentence above would be

Dog likes food.   □

Now we turn to the second interesting property of transformational grammars. Even by considering a simple transformational rule applying to a terminal string (rather than a phrase), we are able to perform some powerful operations.*

7 Lemma. *Let G_1, G_2 be any CFGs. Then there exists a transformational grammar TG, such that TG can perform the intersection of the languages of G_1 and G_2. That is, $L(TG) = (L(CFG_1) \cap L(CFG_2))$.*

PROOF OUTLINE. Let us define a transformational grammar, TG, with a context-free base which has as its only S production $S \rightarrow S_1 \mu S_2 \$$, where S_1 and S_2 are start symbols for grammars G_1 and G_2, respectively. Now let us define a transformation T_1 which will perform the intersection of the two CFGs.

SD:	X	x	μ	X	y	$\$$	w	
	1	2	3	4	5	6	7	$\Rightarrow$
SC:		2	3		5	6	$7+1$	

To ensure that we generate *just* the intersection, we need one more obligatory transformation, T_2, which is $\mu\$ \rightarrow \lambda$. That is,

SD:	x	μ	$\$$	y	
	1	2	3	4	$\Rightarrow$
SC:	1			4	

To give an example of how these two rules work together to perform the intersection operation, consider the following example. Let the two context-free grammars, G_1 and G_2, generate the languages $\{a^n b^n c^m, n, m \geq 1\}$ and $\{a^n b^m c^m, n, m \geq 1\}$, respectively. The transformational grammar described previously, TG, will now generate just the intersection of these two languages,

---

* In the *Aspects*-Model (Chomsky 1965), there is a condition that the deletion be "recoverable," which means that any deletion is performed only under structural identity with some other element. Here this is captured by the copying action of the transformation. For more details, see (Chomsky 1965).

namely $\{a^n b^n c^n, n \geq 1\}$, which, of course, is not context-free. Assume that our grammar has generated the string $aabbcc\mu aabbcc\$$. The transformation T_1 will obligatorily apply to this string until there is no structural description that fits the rule. The rule checks, for each symbol under S_1, whether there is an identical symbol (in the same place) under S_2. If there is, the rule deletes both these symbols and copying it to the right of the end marker $\$$. After all successful applications of T_1, the string is transformed into $\mu\$aabbcc$. Now rule T_2 applies to delete the adjacent markers $\mu\$$. This leaves the string $aabbcc$. Now assume that the grammar generates (before the rules apply) the string $aabbccc\mu aabbbccc\$$. In this case, T_1 will apply until the rule must stop, since there is an unequal number of b's. The second rule cannot apply, however, since the markers are not adjacent. Thus, the string is not generated by the grammar.

Armed with these two results, we now present a theorem concerning the generative capacity of unrestricted transformational grammars.

8 Theorem. *There is an undecidable set S of the natural numbers* **N**, *such that S can be generated by some context-free based transformational grammar.*

PROOF OUTLINE. Recall from Exercise 5.4.10 that we can construct an undecidable set $S \subseteq$ **N** as being the homomorphic image of the intersection of two context-free languages. That is,

$$L = \phi(L_1 \cap L_2)$$

where L_1 and L_2 are context-free languages and ϕ is a homomorphism. The theorem follows by combining this result with the previous two lemmas. □

The relevance of this result should perhaps be explained. If transformational grammars are to be taken as a model for constraining the form of natural languages as we suggested in the previous section, then why should the model generate languages as powerful and unconstrained as undecidable sets? Is there some way to constrain the form of transformations so as to keep the language generated by the grammar recursive, let alone, "natural language-like?"

One might think that by restricting the base rules of the transformational grammar even more tightly than to be context-free, we might keep the resulting language recursive. Unfortunately, however, (Salomaa 1971) gives an even stronger version of the preceding proof with a type-3 base component. It is clear from these results that if we are to restrict the generative power of transformational grammars, it will be necessary to constrain the form of the rules themselves rather than the base. In the next chapter we will review some recent developments in the theory of grammar which speak to the problem of both the generative capacity of the grammar as well as the explanatory adequacy of the linguistic model.

1. Complete the proof of Theorem 5. (*Hint*: Given any FSA (Q, δ, q_0, F) consider the equivalence relation $w_1 \sim w_2$ iff $\delta(q, w_1) = \delta(q, w_2)$ for all $q \in Q$.)

2. What are the Myhill equivalence classes for $L = \{x | x$ has equal number of a's and b's$\}$? Use your result to show that L is not regular.

3. Use Theorem 5 to prove that $L = \{a^k | k = n^2$ for some $n\}$ is not regular.

8.3 Augmented Transition Networks

One of the earliest implementations of a system for parsing a robust portion of natural language was the augmented transition network (ATN), described first in (Woods 1970). This developed out of work on finite-state automata, described in Chapter 2. In this section, we will trace the development of the ATN.

A nondeterministic finite-state acceptor, NDA, characterized as the quadruple (Q, δ, q_0, F) over input alphabet X, can be constructed for a simple fragment of English in the following way. First, consider the state diagram represented next, where PN stands for "proper noun" (e.g., a person's name):

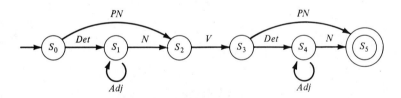

To understand this diagram, let us define some terms and vocabulary. We will distinguish between the actual input vocabulary X, and the *category vocabulary*, C, which is made up of the grammatical categories of natural language, e.g., NP, VP, V. We define the function $CAT: X \rightarrow C$, such that $CAT(x)$ is the defined category of x. For example, if

$$X = \{dog, John, Mary, the, a, loves, saw, white\}$$

and C is:

$$C = \{N, Det, V, Adj, PN\}$$

then we can use the CAT function to define the following relationships:

$$CAT(dog) = N, CAT(John) = PN, CAT(Mary) = PN, CAT(the) = Det,$$

$$CAT(a) = Det, CAT(loves) = V, CAT(saw) = V, CAT(white) = Adj.$$

Next we create a state set $\{S_i\}$ as follows:

S_0 is the initial state;

S_1 is the state in which it is legal for the next word to be the noun of the first NP or its adjectives;

S_2 is the state reached after reading the first NP;

S_3 is the state reached after a legal reading of a verb;

S_4 is the state in which it is legal for the next word to be an adjective or the noun of the next NP; and

S_5 is the "completion" state. Further, $F = \{S_5\}$.

The preceding state diagram corresponds to the following transition function (where $\delta(q, A)$ is shorthand for $\delta(q, x)$ for any x in A, with $A \in CAT$).

$$\delta(S_0, Det) = \{S_1\}$$

$$\delta(S_0, PN) = \{S_2\}$$

$$\delta(S_1, N) = \{S_2\}$$

$$\delta(S_1, Adj) = \{S_1\}$$

$$\delta(S_2, V) = \{S_3\}$$

$$\delta(S_3, Det) = \{S_4\}$$

$$\delta(S_4, N) = \{S_5\}$$

$$\delta(S_4, Adj) = \{S_4\}$$

$$\delta(S_3, PN) = \{S_5\}$$

Finally, $\delta(S, x) = \varnothing$ for any pair (S, x) not listed.

We see that the language accepted by this NDA is

$$L = \{PN \cup Det\,Adj^*N\} \cdot \{V\} \cdot \{PN \cup Det\,Adj^*N\}$$

The sentences accepted by this device include the following:

(1) John loves Mary.
(2) The white cat saw Mary.
(3) Mary loves a cat.

This formalism is unsatisfying as an accepting device for natural language for several reasons. First, notice that the transitions do not capture the effects of constituency, which we showed in Section 8.2 are necessary for an adequate description of natural language. That is, there is no way of stating formally that a path from S_0 to S_2, or from S_3 to S_5, represents a noun phrase. Thus, a finite-state machine fails to capture a significant generalization in the structure of the language. Furthermore, any kind of movement dependencies, such as those discussed in the previous section, are impossible to capture in such a formalism, except at great cost, e.g., by completely replicating most of a network to recreate the effect of movement.

One way of addressing some of the criticism leveled at the NDA model for natural language acceptors is to allow state transitions to refer to non-terminals, or grammatical phrases, as well as to terminal symbols.

The preceding grammar only allows a transition to be made from one state S_i to another state S_j by reading in a single word x of category A such that $\delta(S_i, A) = S_j$. In a *recursive transition network* (RTN) we allow a transition from S_i to S_j to be made by reading in any *string* w such that w is known to belong to a set A for which $\delta(S_i, A) = S_j$. There is a problem here: if A is not a *word*-category, how do we know that w belongs to A? Because w has been accepted by *another* RTN whose job it is to accept strings of A. It is thus clear where the term *recursive* comes in—for the network for recognizing A might call a network which recognizes B which calls ... etc, etc., until we come to a transition which calls on the A-recognizer again. Before looking at recursion, though, let us see how we can use this new idea to restructure our first example:

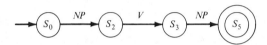

Here V is still an item in C, but the NP transition requires a call to a second RTN, NP.

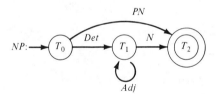

In this RTN every transition is governed by a single word from the corresponding class in CAT.

We apologize to the reader for a slight clash of notation. In the next definition, V is the set of variables, not the category of verbs.

1 Definition. A *recursive transition network*, or RTN, R is given by the following data:

(1) X is a finite set of *terminal* symbols (e.g. the words of a natural language).
(2) C is a finite set of *categories*.
(3) CAT is a map $X \to C$.
(4) V is a set of *nonterminal* symbols.
(5) $S \in V$ is the *start symbol*.

In addition, the RTN includes, for each $v \in V$, an NDA $M_v = \{Q_v, q_v, \delta_v, F_v\}$, where $q_v \in Q_v$, $F_v \subset Q_v$, and $\delta_v: Q_v \times (C \cup V) \to 2^Q$.

We then associate for each $v \in V$ an *acceptance set* T_v, defined recursively as follows: A string w is in T_v just in case it can be partitioned into a sequence $w_1 w_2 \ldots w_n$ such that either (1) $CAT(w_j) = a_j \in C_j$ or (2) $w_j \in T_{a_j}$ for $a_j \in V$, and such that

$$\delta_v^*(q_v, a_1 a_2 \ldots a_n) \in F_v$$

Put differently, a string is accepted by M_v with the transitions from states q to q' in M_v either made with a single x with $CAT(x) = c$ (i.e., the category of x is c) such that $q' \in \delta_v(q, c)$, or by a substring w' which is accepted by some M_u for which $q' \in \delta_v(q, u)$.

Finally, the language $L(R)$ accepted by R is just T_S, the acceptance set for the start symbol S.

2 Theorem. *For every recursive transition network R there is a PDA $P(R)$ which accepts the same language.*

PROOF OUTLINE. Let the state set of $P(R)$ be the disjoint union of the state sets Q_v of the NDAs of R. Let the input set of $P(R)$ be X; and let its set of pushdown symbols be V. At any time, $P(R)$ will be in some state q from Q_v for a particular variable v, and $P(R)$ will then be in the process of verifying whether it is reading a substring that belongs to T_v. When it comes to a symbol, it may continue to process it according to δ_v, or it may choose to check whether it is the first letter of a string in some $M(L_{v'})$. In the former case it applies the appropriate transition sanctioned by (1):

(1) If $P(R)$ is in state q in Q_v, and $q' \in \delta_v(q, c)$, then the transition

$$(q, x, z) \rightarrow (q', z)$$

is allowed for any x with $CAT(x) = c$.

(2) In the second case, it will place v' atop the stack, and go to the initial state $q_{v'}$ of $M_{v'}$: For any q, z, and v' the transition

$$(q, \lambda, z) \rightarrow (q_{v'}, v'z)$$

is allowed.

The idea is that, when $P(R)$ finds a string in $L(M_{v'})$, it may then resume testing for $L(M_v)$ using initial state q' and bearing in mind that it may now make a δ_v-transition from q using input v':

(3) For any q, v, q', and v' with $q \in Q_v$ and $q' \in F_{v'}$, the transition

$$(q', \lambda, qz) \rightarrow (q'', z)$$

is allowed for any $q'' \in \delta_v(q, v')$.

No other transitions are allowed. The reader may verify that $P(R)$ behaves as advertised.   □

The RTN model described is able to overcome many of the criticisms raised against NDAs. Constituency of phrases is captured by allowing grammatical phrases to be treated as states themselves. The problems that arose from unbounded dependencies, however, are still present, since there is no way to relate, for example, a *wh*-sentence with its indicative form. Furthermore, local dependencies, such as subject-verb agreement, are impossible to express.

There is, however, a way to extend the transition structure of an RTN by adding *registers* to the model. A register may hold arbitrary information about the input vocabulary or state sets or tests on any register information. These tests on register values may make reference to values of other registers. The resulting system is called an *augmented transition network* (ATN).

Let us look at two cases involving agreement, one involving the subject and verb of a sentence, the other involving the determiner and head (or main) noun of an *NP*. With the RTN constructed previously there is no way to ensure that sentences like the following are prevented from being accepted as part of the language.

(1) *The *boys is* mischievous.
(2) *A *girls* leave home.

What is needed is some way to check the *number* of the subject NP and somehow make the verb transition conditional on the verb agreeing with this NP feature. The transition along this augmented arc would then be taken only if it satisfies all the normal conditions for an RTN transition as well as the specific conditions mentioned in the register. Thus, the RTN we have above could be made to include subject-verb agreement by augmenting the verb transition as follows:

(i) *Condition*: If the *NP* is *singular* then *V* must be *singular*.
(ii) *Condition*: If the *NP* is *plural* then *V* must be *plural*.

Let us give an explicit example of what a register can contain and how this information is structured. Consider our earlier example, where now we have added a register *R* to each NDA.

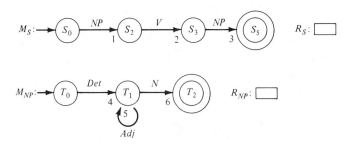

Each register can hold one of three values: *singular, plural,* or *neutral.* Singular does not agree with plural, but neutral agrees with everything. The transitions have been numbered for ease of reference. Each NDA is *augmented* as follows:

Transitions for M_S, where w is a phrase (or network) variable, and x is a word (or arc) variable:

Transition 1. $\delta_S(S_0, w) = S_2$ if $w \in T_{NP}$. When the transition is made, perform the assignment $R_S := n$, where n is the value set in R_{NP} by the acceptance of w by M_{NP}.

Transition 2. $\delta_S(S_2, x) = S_3$ if $CAT(x) = v$ and the number of x agrees with that stored in R_S.

Transition 3. $\delta_S(S_3, w) = S_5$ of $w \in T_{NP}$.

Transition for M_{NP}:

Transition 4. $\delta_{NP}(T_0, x) = T_1$ if $CAT(x) = Det$. In making the transition, $R_{NP} :=$ the number of x.

Transition 5. $\delta_{NP}(T_1, x) = T_1$ if $CAT(x) = Adj$, and the number of x agrees with that stored in R_{NP}.

Transition 6. $\delta_{NP}(T_2, x) = T_1$ if $CAT(x) = N$, and the number of x agrees with that stored in R_{NP}.

On exit from T_2, a string w in NP carries the number stored in R_{NP} at that time.

Let us now consider how *long-distance dependencies* can be handled in an ATN formalism. Recall that neither FSAs nor RTNs were able to parse the following kinds of sentences involving *wh*-Movement.

(1) What does John love?
(2) Who loves Mary?

Without replicating the entire arc path for the declarative form with the addition of an initial *wh*-element, we are currently unable to handle such sentences. There are several alternative ways of analyzing the dependencies between the *wh*-phrase and the empty element in these sentences. We will discuss only one, the *global register*.

Let us allow some registers to be associated with the whole sentence. Such registers are called global registers because they can be modified and accessed globally, by any transition. For example, to parse sentences (1) and (2), we would modify our ATN to include the following types of transitions. We add a global register R_{wh} which can be empty, hold *wh*, or hold *do*. We add to M_{NP} a new transition:

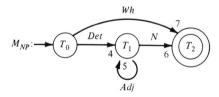

Transition 7. $\delta_{NP}(T_0, x) = T_2$ if $CAT(x) = Wh$. In making the transition, $R_{wh} := x$, and $R_{NP} := neutral$.

We add to M_S two extra transitions:

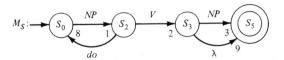

Transition 8. $\delta_M(S_2, x) = S_0$ if $CAT(x) = do$ and $R_{wh} = wh$. In making the transition, $R_{wh} := do$.
Transition 9. $\delta_M(S_3, \lambda) = S_5$ if $R_{wh} = do$.

We also add to Transition 3 the condition $R_{wh} \neq do$. Note this structure accepts (1), (2), and (4), while rejecting (3).

(3) *Who likes?
(4) Who likes who?

With the addition of registers, then, we have overcome the difficulties encountered by the more restricted models of parsing, FSAs and RTNs. However, the ability to augment a transition with arbitrary tests and computations gives an ATN the computational power of Turing machines. In other words, while the ATN gives us much of the descriptive power we need to represent grammars for natural languages, it also lets us describe processors with structures quite unlike those of natural language processors.

3 Theorem. *For each Turing machine T, there is an ATN which accepts precisely the strings accepted by T.*

PROOF. See Exercise 2. ☐

EXERCISES FOR SECTION 8.3

1. Consider the final ATN from Section 8.3. Which of the following sentences will it presently accept?
   a. The man saw the cat with the telescope.
   b. Mary eats soap.
   c. What did Mary see Bill with?
   If it does not accept one of these sentences, indicate what changes or additions have to be made to the *ATN* to accept them.

2. Prove Theorem 8.3.3. (*Hint*: The idea is similar to the proof of *RTN* equivalence to *PDA*s. Show that for every two-stack PDA there is an ATN which accepts the same language.) (See Exercise 5.1.3.)

CHAPTER 9

Recent Approaches to Linguistic Theory

Because unconstrained transformational grammars are capable of generating undecidable sets—apparently more power than is necessary for natural languages—much recent research within linguistics has focused on constraining the power of grammars. Several approaches to this task have been taken, and we will examine two in detail. First, however, in 9.1 we will review some current developments in linguistic theory that are common to both approaches to be examined. Then, in Section 9.2 we describe the framework of *generalized phrase structure grammar* (GPSG). Finally, in Section 9.3 we look at the theory of *government and binding* (GB), the current version of transformational grammar.

9.1 Beyond Unrestricted Phrase Structure Formalisms

In the previous chapter we viewed the grammar for a natural language as consisting of a phrase structure component and a transformational component. We did not discuss the other important aspects of a grammar, such as the semantic interpretation for a syntactic expression, and the structure of the lexicon (that is, how the dictionary for a language is structured). For the past 15 years, however, these areas have been studied extensively and the shape of linguistic theory has changed as a result. Some of the more significant changes involved have been to:

1. Enrich the structure of the lexicon.
2. Restrict the power of the context-free base grammar.

3. Reduce the power of transformations.
4. Examine the form of semantic interpretation.

Let us now review briefly each of these areas.

In our discussion of transformations in the previous chapter, we introduced a few of the transformations that were employed by grammarians to account for the wide range of data in natural languages. To give a taste for some of the less "structure-preserving" rules (that is, rules that change the structure of the string in a significant way), consider the rule for deriving *complex nominals*. This rule takes a sentence such as (a) as input and derives a noun phrase (b), the nominalization of the sentence.

(a) John destroyed the book.
(b) John's destruction of the book.

There is good motivation for relating the two words *destroy* and *destruction*, since they have the same "thematic" structure (i.e., their meaning is similar). In a transformational model such as that outlined in the previous chapter, but which we are about to move beyond, there is no way for such items to be related except by transformations. Without giving the rule in detail, it is clear that such transformations would involve four major parts:

(1) The entire phrase must change categories from *S* to *NP*.
(2) A *PP* must be constructed with *of* as the preposition.
(3) *John* must change to *John's*.
(4) The verb *destroy* must change to the noun *destruction*.

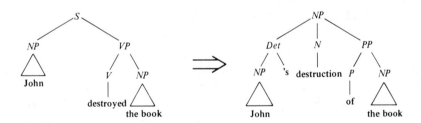

It should be clear that these are just the sort of unconstrained transformational rules that gave us Theorem 8.2.8. Furthermore, the ability to create structure that is already motivated and needed for other parts of the language (that is, the resulting *PP* which is built), as well as the power to change categories (from S to *NP* in this case), makes the recognition problem for a parser very difficult.

Rather than use transformations to relate structures (such as those in (a) and (b)) which bear semantic similarities, we can generate both forms using the phrase structure rules of the grammar and relate the words *destroy* and *destruction* lexically. With this alternative approach, no transformation is needed to derive the nominal form.

The transformational rule deriving the nominal form was meant to capture the intuition that both the verb and nominal form take the same complement, *the book*, and that it functions as an object in both structures. We should note here that linguists use the term *complement* to stand for any kind of object, e.g., the direct object of a verb or the noun phrase that functions as the object of a preposition.

Without transformations we can identify the phrase containing the noun and its complement with the phrase containing the verb and its complement as forming an equivalence class. That is, the VP in (a) forms an equivalence class with the phrase, *destruction of the book*. Chomsky (Chomsky 1970) was the first to develop a theory of phrase structure based on equivalence relations between constituents. His analysis has greatly influenced the direction of linguistic research.

By borrowing the concept of a *feature vector* from phonology, we can think of primitive categories in the language as decomposable into smaller units. To make this clearer, let us think of categories and nonterminals in a language as being *record structures*. Associated with each record is a collection of named fields, each carrying a specific value. For example, imagine that we have two fields called *noun* and *verb* which can have value + or − (equivalent to the more familiar Boolean values 1 and 0).

Such structures give rise to the four major classes of speech, or word types, in the language. A shorthand representation for these structures is the following:

$N: [+N], [-V]$ (Noun)
$V: [-N], [+V]$ (Verb)
$A: [+N], [+V]$ (Adjective)
$P: [-N], [-V]$ (Preposition)

It is clear why a noun carries the feature set $[+N]$ and $[-V]$, why a verb is marked $[+V]$ and $[-N]$. Perhaps not so obvious is why adjectives and prepositions are marked as they are. This convention, however, is motivated by some intuitive observations about natural languages. Consider the features for adjective first. In many languages an adjective will "agree" with the noun it modifies, just as the subject of a sentence agrees with the verb. For example, in German, there is obligatory agreement between adjective and noun within an *NP*.

(1) die roten Rosen
(2) the red roses

Thus, just like nouns and verbs, adjectives show agreement. Of course, the determiner in (1) also agrees, so this is not a unique property that adjectives share with nouns and verbs. There is another indication that adjectives have nounlike properties, however, and this can be illustrated with an example from Turkish. Turkish is a richly inflected language, where suffixes mark many of the features and functions expressed by full words in a language such as

English. An interesting feature of Turkish (and many other languages) is that adjectives can be used as nouns. For example, from the adjectives in (1) and (2), it is possible to derive nouns with the use of suffixes.

(1) *genç* "young"
(2) *hasta* "ill"
(3) *bir genç* "a young person"
(4) *gençler* "the young"
(5) *bir hasta* "a sick person, patient"

Thus, adjectives do seem to be able to act as nouns. Adjectives also act as verbs, however, in that they can carry verbal suffixes as well. For example, the Turkish adjective *tembel*, "lazy," can be turned into the verbal form *tembelyim*, "you are lazy," by a suffix meaning *to be*. Given these two types of behavior, it makes sense to label the category adjective $[+N], [+V]$.

What is there to say about the category of prepositions? It is the *lack* of any similarity to the other categories that motivates our assigning it the feature $[-N], [-V]$.

Now let us introduce another feature (or record structure field) for categories, the feature *BAR*, which determines the *level* of a phrase. An example will clarify the exact meaning of this feature. We saw above that both the verb *destroy* and the noun *destruction* allow an object or complement (namely, *the book*).

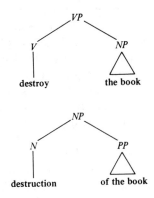

The presence of *of* in the *NP* case will not concern us here. We can think of it as being inserted to assign case to the object *NP* since nouns cannot assign case as verbs do.

Let us say that the phrase which contains both the *N* and its complement (the entire tree rooted at *NP* in the second illustration) is a *projection* of the features of *N*, where we can think of this node as *inheriting* these particular features from below. To distinguish between the node *N* and the node dominating it, we introduce a new feature, *BAR*. *N* has feature $BAR = 0$, whereas the node dominating it has $BAR = 1$. This is illustrated next, where we replace

NP with a complex expression, $\{[+N],[-V],\dots\}$, and so on, for the other nodes.

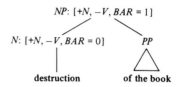

Similarly, the *V* and phrase containing it are distinguished by the value of the feature *BAR*.

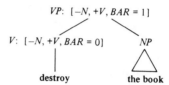

We can think of the number of "bars" associated with a node as the *level* of that phrase. For example, the phrase in the preceding figure, call it $\bar{V}$, is of "level 1" and the corresponding phrase with the noun *destruction* and its complement is an $\bar{N}$, also of "level 1." We define an equivalence relation, "having n bars." Then the equivalence class $\bar{X} = \{\bar{N}, \bar{V}, \bar{A}, \bar{P}\}$ is the class of phrases containing a lexical item and its complement. These classes correspond to words that are two-place predicates, for example, $X(y, z)$, where X is the predicate and y and z are the arguments. If this $\bar{X}$ is to be a genuine linguistic construct, adjectives and prepositions must take complements as well. This is indeed so, since there are adjectives which have two arguments such as *afraid*, e.g., "John is *afraid of* snakes," *afraid(John, snakes)*, as well as two-place prepositions such as *under*, e.g., "The cat is *under* the table," *under(cat, table)*. This is in contrast to one-place predicate adjectives such as *red* and *sick*, as in the sentence "The cat is *sick*." The similarity in syntactic structure of $\{\bar{N}, \bar{V}, \bar{A}, \bar{P}\}$ is illustrated by the following examples.

(a) the $[_{\bar{N}} [_N \text{painting}]$ of Bill$]$

(b) John [$_{\bar{V}}$ [$_V$ painted] Bill]

(c) John is [$_{\bar{A}}$ [$_A$ afraid] of snakes]

(d) The cat is [$_{\bar{P}}$ [$_P$ under] the table]

Because all possible categories defined by the two-feature system described permit complements of one sort or another, we can introduce the variable X to refer to any of the four categories and generalize these four structures to one phrase structure rule:

$$\bar{X} \rightarrow X \; Comp$$

In such a rule, X is the *head* of the phrase and *Comp* is the complement or *foot* of the phrase. This rule is a collapsing of the following four rules.

$$\bar{V} \rightarrow V \; Comp$$
$$\bar{N} \rightarrow N \; Comp$$
$$\bar{A} \rightarrow A \; Comp$$
$$\bar{P} \rightarrow P \; Comp$$

As is clear from the examples, *Comp* is really *Comp$_X$*, i.e., it depends on the context of X chosen. In addition to a general notion of complement, Chomsky proposed that there is a cross-categorial notion of *specifier*, or *Spec*, which is the sister to the phrase $\bar{X}$. Intuitively, this is a phrase which acts to specify or make definite in some way, the phrase to which it is sister. For example, in the *NP*, *the green ball*, the determiner *the* is a specifier indicating the definiteness of the object.

The $\bar{X}$ together with its specifier make up what is called $\bar{\bar{X}}$.

$$\bar{\bar{X}} \rightarrow Spec\ \bar{X}$$

where again *Spec* is shorthand for $Spec_X$. This in turn is an abbreviation for the following four rules.

$$\bar{\bar{V}} \rightarrow Spec\ \bar{V}$$
$$\bar{\bar{N}} \rightarrow Spec\ \bar{N}$$
$$\bar{\bar{A}} \rightarrow Spec\ \bar{A}$$
$$\bar{\bar{P}} \rightarrow Spec\ \bar{P}$$

Although the notation may be new, the structures here are familiar. The rule for $\bar{\bar{N}}$, for example, is a slightly different version of the following *NP* rule.

$$NP \rightarrow Det\ A\ N\ PP$$

As mentioned, the specifier for an *NP* is usually the determiner. In the *NP* "the destruction of the book," this gives rise to the structure shown (where the Adjective node *A* is not shown in this example).

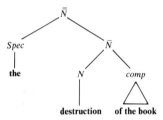

Similarly, the rule for $\bar{\bar{V}}$ is the *VP* rule, where the *Spec* is the position for auxiliary verbs, such as "should," "have," "has," etc. This accords with our intuitions, since these auxiliary verbs act to further specify the activity denoted by the verb. An example of the *VP* "has destroyed the book" from the sentence

John has destroyed the book.

is illustrated now.

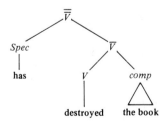

The two rules,

$$\bar{\bar{X}} \to Spec\ \bar{X}$$

$$\bar{X} \to X\ Comp$$

are the essential components of what is called $\bar{X}$-theory. This theory states that the base rules (the context-free grammar) for all natural languages must conform to the following highly restricted form of rewrite rules:

$$X^n \to (Y^n \ldots)X^{n-1}(Z^n \ldots)$$

where X, Y, and Z range over values of the features $[\pm N]$ and $[\pm V]$ and the superscript indicates the bar level of the phrase. The node X^0 we will call the *head* of the phrase, since it is through the path beginning at the leaf X^0 that the features are inherited.

The important of $\bar{X}$-theory for linguistics is that the nature of the base, the underlying phrase structure grammar for a language, is very restricted in its form. All productions must conform to this template. For a further study of $\bar{X}$-theory, see Jackendoff (1977) and van Riemsdijk and Williams (1986).

THE STRUCTURE OF THE LEXICON

Another important area that has been intensively investigated in modern linguistics is the lexicon. Common to both theories that we will examine in the next sections is the idea that a lexical entry (i.e., a word) contains much more information than was initially used in transformational grammar. One such enrichment involves the feature *CASE*, which is associated with verbs, prepositions, and the morpheme for expressing the tense of the verb, which is the marker on an inflected verb that agrees with the subject noun phrase. Tense assigns *nominative* case to the subject, the verb assigns *accusative* to its direct object, and a preposition assigns *oblique* case to its object.

In modern linguistic theory all $[-N]$ categories are case assigners, since P is $[-N, -V]$, V is $[-N, +V]$. Moreover, today many linguists believe that tense itself is a category that holds features and is considered to be $[-N]$, since it attaches to verbal elements, and in many languages is in fact a verb. An NP must be assigned the correct case which in English is determined by the position this NP occupies in a sentence. We can see this with pronouns in sentences such as those in (1).

(1) a. *He* saw Mary.
   b. Mary saw *him*.
   c. **Him* saw Mary.
   d. *Mary saw *he*.

The pronoun *he* in (1a) is in the nominative case, and substituting with the accusative pronoun *him* results in an ungrammatical sentence (1c). The same thing is happening with the accusative pronoun in (1b).

By stating that all NPs in a sentence must receive some case (from some case assigner), we impose a partial correspondence between the number of verbs and prepositions in a sentence and the number of *NP*s. For example, the following sentence in (a) is ungrammatical because there is no case assigned to the subject *NP* (since the infinitive marker *to* doesn't assign case). The sentence in (b) is grammatical, however, since the tense marker *-s* assigns nominative case to *John*.

(a) *John to eat the fish.
(b) John eats the fish.

Another distinction that the feature *CASE* can make is between verbs and adjectives. Recall from the previous chapter that the passive rule applies to transitive verbs, verbs such as *eat*, which assign case to their complements (*the fish* in (a) and (b) is assigned accusative case, for example). But when transitive verbs appear in the passive form (i.e., *eaten*),

(c) The fish$_i$ was eaten $[e]_i$.

they are no longer able to assign accusative case to an object—the subject in (c) was originally in object position, hence the trace. We can think of the operation of adding one of these suffixes, such as *-en* or *-ed*, as turning the verb into an adjective, which assigns no case. This is important for several reasons. First, the lack of case assigned by the passive form of the verb means that an *NP* in object position in a sentence will not receive case from this word—although a trace appears in the object position in (c), a full *NP* is not permitted. Such is the situation in (d), which is ungrammatical for this reason.

(d) *The fish was [eaten]$_{[+N, +V, -CASE]}$ [the bug].

Second, it points to the significant role that suffixes and prefixes (affixes) can play in a language. The study of this area, called *morphology*, deals with how affixes affect the case-assigning properties of verbs, as well as how they can affect the argument structure of a word.

In this section we have outlined some of the basic changes in the development of the transformational model. By far the most important change is the restriction placed on the form of phrase structure rules defined by $\bar{X}$-theory. In the next two sections we will examine two different approaches that make use of this system.

9.2 Generalized Phrase Structure Grammars

In this section we will examine the framework of *generalized phrase structure grammar* (GPSG), as outlined in (Gazdar et al. 1985). Gazdar (1981) claimed that the overgeneration resulting from the unrestricted transformational component of a transformational grammar could be avoided by simply eliminating transformations from the grammar entirely. That is, it is possible on this view to capture the generalizations made by transformations within a purely context-free grammar. This could be achieved, he argued, with the use of special features on nodes, which would allow long-distance dependencies in sentences of a language without movement rules.

With a basic understanding of the notions of features and $\bar{X}$-theory from Section 9.1, let us see how GPSG builds on Chomsky's original proposals to incorporate mechanisms that recreate the *effects* of transformations within a context-free formalism. The essential components of GPSG are given next:

(1) A theory of features
(2) A grammatical rule format
(3) The properties of metarules, and
(4) A theory of feature instantiation principles

First to item (2). We assume the standard interpretation of context-free phrase structure rules, where the form is

$$A \rightarrow BC$$

Now let us turn to item (1), the theory of features in GPSG, since it is so central to the theory, and see how it differs from that proposed by Chomsky. In the original theory of $\bar{X}$-syntax, all features were of one type: *Boolean*. That is, the value of each feature was either $+$ or $-$. In GPSG there are two types of feature:

(1) *atom-valued*: including Boolean values as well as symbols such as $[-INF]$ (finite, an inflected verb, e.g., "eats"), $[-INV]$ (inverted, e.g., *subject-auxiliary inversion*, as in *Is John sick?*), and $[+INF]$ (infinitival, e.g., "to eat").*
(2) *category-valued*: where the value of a feature is something like a non-terminal symbol (which is itself a feature specification). *SUBCAT* is such a feature that identifies the complement of the verb. One of the two basic examples of this feature type is *SLASH*. If we have a normal transitive verb phrase, *VP*, such as "hit the floor", then $VP[SLASH = NP]$ (or in an abbreviated notation, VP/NP) will represent this *VP* when it has an *NP* missing (a "*VP* with an *NP* gap"); for example, "hit $[e]$", as in "Who did John hit?". Broadly speaking, VP/NP will stand for a *VP* with a missing *NP*; an S/NP is a sentence with a missing *NP*. In general, the *SLASH* feature plays the role that traces play in transformational grammar.

To handle *wh*-questions, we need to introduce another feature besides *SLASH* in order to encode the "questionlike" nature of these sentences. That is, we introduce a feature, $+WH$, which differentiates the following *NP*s:

---

* For our purposes, we are collapsing two separate features, *INF* and *FIN* into one. See (Gazdar et al. 1985).

(a) $-WH$ [the man]
(b) $+WH$ [which man]
(c) $-WH$ [John]
(d) $+WH$ [who]

There are two additional notions that are necessary in order to understand the feature system: these are *extension* and *unification*, both definable over categories (as *feature specifications*). Basically, the *extension of a feature specification* is a larger feature specification containing it. As an example, consider the feature specification $\{[+N],[+V]\}$, namely, the category A, adjective. A legal extension of this category is $\{[+N],[+V],[+PRED]\}$, which is an adjective in a predicative position. Here the adjective acts as the predicate of the NP in subject position, unlike its role as a modifier when it appears within the NP. This is illustrated next.

$$\text{Mary is } [_{\{[+N],[+V],[+PRED]\}} \text{ intelligent}]$$

Unification is also defined for categories, and it is similar to the *set union* operation. Take as an example the two feature specifications $\{[+V], [+PRED]\}$ and $\{[-N],[+V]\}$. The unification of the two is $\{[+V], [+PRED],[-N]\}$. In the case where features contradict each other, the unification operation is undefined.

The third important aspect of GPSG is the notion of *metarules*. Essentially, a metarule can be thought of as a function from lexical rules to lexical rules. A metarule performs much the same function that certain transformations did in our earlier discussion of transformational grammar, but metarules act by making related phrase structure rules available in the grammar, rather than by postulating a two-step process in which one form—deep structure—is changes to a second form—surface structure. Recall from Section 8.2 that, in earlier treatments of transformational grammar, the active and passive forms of a sentence were related by a transformation, the *passive rule*. Sentence (2) was seen as having undergone this rule, but not (1).

(1) John washes the car.
(2) The car is washed by John.

The passive rule allowed us to capture important generalizations, including what role the subject plays in a passive structure, and how this role relates to the object in an active form. Although we could write rules to generate (2) directly, we argued against this approach earlier as failing to capture these generalizations. But the metarule mechanism in GPSG seems to capture this observation. The metarule for passive will look like the following.

PASSIVE METARULE

$$VP \rightarrow W\ NP \Rightarrow VP[PASSIVE] \rightarrow W(PP[+by])$$

That is, for every context-free rule introducing a VP as an NP and some variable number of constituents (including the verb) indicated by W, there is another context-free rule where the *VP* is marked with the atom-valued feature $[+PASSIVE]$, the *NP* present in the active form is missing, and an optimal PP is introduced, marked with the atom-valued feature $[by]$, which means "selects preposition *by*." The presence of the varying parameter *W* makes this a metarule; standard rewrite rules are produced when *W* is instantiated. For example, together with other principles of the theory, this rule says that for a *VP* such as

(1) [$_{VP}$washes the car]

    there is also a *VP* such as
(2) [$_{VP}$washed (by *NP*)]

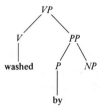

    Notice that the passive metarule makes no reference to the subject of the sentence. This is because the semantics for the verb will be different for different instantiations of the metarule. Thus, what was captured by a transformation (the relation between active and passive forms) is now handled by a metarule and the associated semantic rule; compare (Gazdar et al. 1985) for details.

    Metarules will, in general, capture generalizations made by *local* transformations in a transformational grammar. In order to understand how long-distance dependencies are treated in GPSG, we must understand the fourth important notion of the theory, that of *feature instantiation principles*. Let us turn to this next.

    Recall from $\bar{X}$-theory that a phrase structure rule specifies that *one* category is the *head* of the phrase. Recall that the *head* of a phrase is the category-defining element in the phrase (i.e., *N* in $\bar{N}$ and *NP*). The *foot* is generally the complement of the phrase (i.e., *Comp* in *NP* → *N Comp*). We will distinguish two sets of features for characterizing various grammatical properties:

(i) *HEAD* features $= \{N, V, PLURAL, PERSON, PAST, BAR, \ldots\}$
(ii) *FOOT* features $= \{SLASH, WH\}$.

The features in (i) are called head features because they are properties of the head elements of rules. These take as their values + or −. Foot features are the features that encode the more complex information about movement of *wh*-phrases and NPs. They take on categories as values.

The head shares many features with its mother node, i.e., the *HEAD* features. Thus, if a verb is specified as [+*FIN*] (finite) and [+*SNG*] (singular), the *VP* or $\bar{V}$ dominating it will inherit these features. Informally, we state the behavior of heads and the features associated with them next.

THE HEAD FEATURE CONVENTION

> The *HEAD* features of a daughter node must be identical to
> the *HEAD* features of the mother.

The second set of features, *FOOT*, encode grammatical information on a node that a constituent is missing from within the phrase. These features allow us to capture the effects of "movement" in a transformational grammar. In order to see how, we first state the instantiation principle associated with these features.

THE FOOT FEATURE PRINCIPLE

> *FOOT* features instantiated on a mother category in a tree
> must be identical to the unification of the instantiated *FOOT*
> feature specifications in all its daughters.

To illustrate this, let us return to an example of a *wh*-question that we encountered in Section 8.1.

(a) *Who* drives a Honda?
(b) *What* does John drive *e*?

In our original treatment of such constructions we introduced a transformational rule which acted to move the wh-phrase *who* or *what* from the deep structure position (marked with a "trace" *e*) to the front of the sentence. Within GPSG we can generate this same sentence without the application of a transformation. Let us see how this is accomplished, ignoring details of subject-auxiliary inversion for the time being.

The basic idea is to encode the "movement" information on the node of a tree directly and pass this information up and down the tree with the use of features. First, consider the rule for generating a simple sentence, such as the following

(c) John drives a Honda.

where *TV* stands for transitive verb, and the *SUBCAT* feature has the value

NP (that is, it takes an NP object). Thus,

$$TV = \{[+V], [-N], [SUBCAT = NP]\}$$

The rules necessary to build the sentence in (c) are

$$S \rightarrow NP \; VP$$

$$VP \rightarrow TV \; NP$$

Now, in order to generate a *wh*-movement sentence, such as (b), we assign the value NP to the feature $SLASH$ on the VP node. This indicates that there is a constituent missing from that phrase. Thus, the use of $SLASH$ captures the same generalization that the movement rule did in transformational grammar. In GPSG, a rule of the form $VP \rightarrow V \; NP$ implies a rule of the form

$$VP/NP \rightarrow V \; NP/NP$$

in accordance with the foot feature principle. Similarly, the rule $S \rightarrow NP \; VP$ allows two other rules,

$$S/NP \rightarrow NP \; VP/NP$$

$$S/NP \rightarrow NP/NP \; VP$$

where VP/NP, etc., is a shorthand notation for $VP[SLASH \; NP]$ and S/NP is shorthand for $S[SLASH \; NP]$. What this first rule says is that a VP with a missing NP, called an VP/NP, yields a "trace" by the marking NP/NP. The second set of rules states that a sentence with a missing NP rewrites to either an NP with a VP containing a missing NP or a trace with a VP.

Using the two features WH and $SLASH$, we are able to account for the *wh*-questions illustrated earlier. Assume that the rules for expanding the sentence are given as follows.

$$S \rightarrow NP \; VP$$

$$S \rightarrow NP \; S/NP$$

When the feature $[+WH]$ is added to S, we get the following two structures, after applying the foot feature principle, which transmits the feature information downward in the tree. It should be noted that WH is a feature that cannot cooccur with the feature $SLASH$.

These trees correspond to the sentences

(a) *Who* drives a Honda? and *What* drives a Honda?
(b) *What* does John drive? and *Who* does John drive?

respectively.

The full tree structure are illustrated in the following:

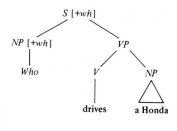

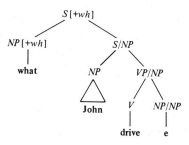

With the power of features introduced by GPSG we are able to represent unbounded dependencies (such as *wh*-movement) in a context-free formalism. Given this fact, we can modify our algorithms for parsing context-free languages accordingly and parse natural languages as well. Schieber (1984) discusses just such a parser, and many variations have been developed. The reader should see (Schieber 1986) for further discussion on parsing in a framework such as GPSG.

9.3 Government and Binding Theory

In this section we will describe another approach taken to constrain the generative power of grammars for natural language. The theory is called *government and binding* (GB) and can be viewed as the direct outgrowth of transformational grammar as described in Chapter 8. The current theory differs from its precursor in many important respects, however, as we shall see. The term *government* refers to a very local structural relation obtaining between two nodes. *Binding* is the term used to refer to antecedent-variable relations. It is similar to the concept in quantified logic, in which a quantifier binds a variable, as when $\forall x$ binds all occurrences of the variable x within its scope.

GB theory and transformational grammar (TG) differ in two important ways:

(1) In GB, the logical form (or semantics) of a sentence is considered to be just as important as the surface structure of a sentence (the words that we actually *hear*).

(2) Most of the transformations from TG have been collapsed into one general operation, called *move* α, which relates the unmoved structure with the moved structure.

In classical TG as described in our previous chapter, a sentence such as (1) was generated using the phrase-structure rules in (2).

(1) The man hit the table.

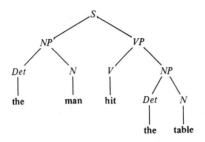

(2) $S \to NP\ VP,\ NP \to Det\ N,\ VP \to V\ NP$.

In addition to these rules in the phrase structure component of the grammar, we specified a certain amount of lexical information for the verb *hit*. We called this a *subcategorization frame*. This is shown in (3).

(3) *hit*: $V, \langle NP \rangle, (Agent, Patient)$.

where the thematic relations are also specified. The *complement slot*, or subcategory, of this verb ($\langle NP \rangle$) specifies an adjacent *NP* as the complement of *hit*. Notice that this is exactly the kind of information that was specified by the phrase structure (PS) rule itself. In an effort to get rid of this kind of redundancy in the grammar, GB has chosen to reduce the number of specific PS-rules, such as the *VP* rule, and let the work be done by the lexicon (*i.e.*, subcategorization) and the principles of $\bar{X}$-theory. Thus, since the lexical entry for *hit* specifies an *NP*, this means that the *NP* is its complement. Now recall from our discussion of $\bar{X}$-theory that we have the two general rule templates.

$$\bar{\bar{X}} \to Spec\ \bar{X}$$

$$\bar{X} \to X\ Comp$$

The *X* here is any lexical category, such as *verb*, and *hit* will appear in an instantiation of this rule, given in (4), where the *Comp* is spelled out as *NP*, as selected by the rule in (3).

(4) $\bar{V} \to V\ NP$

This builds the tree in (5), using the lexical information that $\langle NP \rangle$ is the complement of *V*, and the *specific* phrase structure rule $\bar{V} \to V\ NP$ is never explicitly mentioned.

(5)

Now consider a more complicated *VP*, one that has two arguments in the VP.

(6) John [$_{VP}$has put [$_{NP}$the book][$_{PP}$on the table]].

Rather than listing a rule that generates this VP, all we need do is (a) look at the subcategorization (the complements slot(s)) of the verb, and (b) match it into a template specified by $\bar{X}$-theory. In this case, the subcategorization for the verb *put* is given in (7).

(7) *put*: ⟨*V*⟩, ⟨*NP PP*⟩, (*Agent, Patient, Location*)

Once again, matching up this entry with the appropriate template provided by $\bar{X}$-theory gives the tree in (8):

(8)

The subcategorization information that is encoded in a lexical item is *instantiated* in the manner just illustrated. Also included as lexical information are the *case-assigning* properties of the verb as well as the *thematic* relations (or case roles) it assigns (cf. Section 9.2 for discussion).

The other major shift in emphasis taken in the development of GB has been to examine the logical form (LF) of a sentence as well as its syntactic form. It is not enough to give a structural description of a sentence; one must also give a logical interpretation showing what the sentence means. In earlier TG the meaning of a sentence was derived entirely from the deep structure. Transformations then applied to these structures without changing the meaning to give the surface structure. In this model then, the meanings of (9a) and (9b) are the same, since the interpretation is done at deep structure.

(9) a. John ate the pizza.
     b. The pizza was eaten by John.

In the current view of grammar, the interpretation of a sentence is dependent on the structure that we hear (the *S-structure* in current terms) and the behavior of particular phrase types. The movement of an *NP* through passivization has the effect of leaving a *trace*:

(9c) [The pizza] was eaten [*e*] by John.

The *NP* "the pizza" can recover the grammatical relation information from the trace, which is in object position. That is, we know that it is the *Theme* of the action in the sentence. Since movement rules leave a trace of the moved element, semantic interpretation is possible from the *S-structure* of the sentence.

Let us examine another example. Consider the sentences in (10) where multiple quantifiers are involved.

(10) a. Everyone in this room speaks two languages.
     b. Every Russian loves some woman.

Sentence (10a) can be paraphrased as "For each person in this room, it is the case that he speaks two languages." There is another reading, however, which can be paraphrased as "There are two languages such that everyone in this room speaks them." The same sort of ambiguity exists for sentence (10b). It may be that every Russian is involved with a different woman, or that there is one woman, Olga, whom every Russian loves. The readings of (10a) can be represented in logic notation as

(11) a. $\forall(x)\exists 2(y)\,[\textit{in-this-room}(x) \to \textit{speak}(x, y)]$
     b. $\exists 2(y)\forall(x)\,[\textit{in-this-room}(x) \to \textit{speak}(x, y)]$

GB responds, essentially, by incorporating such structures into linguistic theory. In addition to the underlying level, *D-structure*, and the surface structure, *S-structure*, we introduce another level, the logical interpretation of a sentence, and call it *logical form* (LF). Thus the surface structure (10a) has the two distinct logical forms (11a) and (11b).

First, let us examine the S-structure of sentence (10a).

(12)

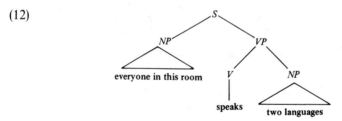

Both *NP*s in this sentence are what logicians and linguists term *quantifier phrases*, since they denote sets of individuals of some particular quantity.

In order to express the meaning of such quantifier phrases as quantifier-variable bindings, GB theory adopts the convention that logicians have had since Frege's time that quantified expressions are in normal form when they appear at the front of an expression and bind a variable. GB proposes that all quantifier phrases are *moved* to the front of the sentence, to a position called C. This movement leaves a trace just as *wh*-movement and *NP*-movement do.

The phrase structure rule that introduces the C node is

(13) $\bar{C} \to CS$.

We can think of this as just another instantiation of the $\bar{X}$-template, $\bar{X} \rightarrow X$ *Comp*, where $\bar{C}$ is a projection of the head C, and S is the complement of C.

Now let us return to the sentences in (10). Even though we *hear* the S-structure, we can *interpret* the sentence (10a) to mean either (11a) or (11b). One of the logical forms of sentence (10a), (11a) will have the structure in tree form given in (14).

(14)

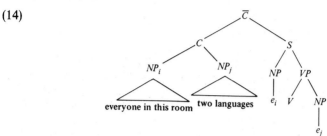

Here the universal quantifier takes wider scope; that is, it is outside the scope of the other quantifier, the existential. But because the *move quantifier phrase* (Move-QP) rule applies blindly, there is another equally valid structure in which the two QPs appear in reverse order:

(15)

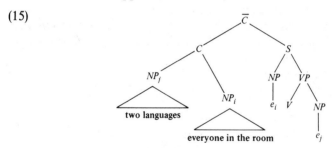

In conclusion then, we can use the unrestricted nature of the Move-QP rule to explain the inherent ambiguities in the sentences in (10).

WH-MOVEMENT REVISITED

Earlier we discussed the transformation *wh*-movement which transformed the D-structure in (1a) to the well-formed S-structure in (1b).

(1) a. John has visited *who*
     b. *Who*$_i$ has John visited e_i

In current GB theory, quantifier movement and *wh*-movement are seen as the same rule, but operating on different *levels* in the grammar. Let us clarify this notion a bit. Imagine that the grammar contains levels of interpretation, which describe different aspects of a sentence. The model in GB looks something like that in (2).

(2) *D-Structure*

⇓

S-Structure

⇓

Logical Form

Wh-movement applies to a D-structure and derives an S-structure. Thus, in English, the sentence that we *hear* contains the *wh*-phrase at the beginning. QP-movement, on the other hand, applies to S-structure and derives the logical form. This is an interpreted level and is not that form which we hear.

The formulation of *wh*-movement in GB is quite simple and looks just like QP-movement.

(3) *Quantifier phrase movement*: Move any QP_i to the nearest C node and leave a trace e_i of the movement.

(4) *Wh-movement*: Repeatedly move any *wh*-phrase WhP_i, to the nearest C node, until reaching a C node marked $[+WH]$, leaving a trace e_i of each movement.

Let us give an example to illustrate more clearly how this rule works in the current theory. Remember that S can refer to either a subsentence or the overall S we are examining. Consider the sentence in (5a), where (5b) gives the detailed D-structure. Since the entire sentence is a question, we mark the first C with a $[+WH]$.

(5) a. *Who* does John believe Mary likes?

   b. $[_{\bar{c}} [_{c[+WH]}] [_s \text{John believes} [_{\bar{c}}[_c] [_s \text{Mary likes who}]]]]$

Now the rule applies, giving the intermediate structure in

(6) $[_{\bar{c}} [_{c[+WH]}] [_s \text{John believes} [_{\bar{c}}[_c \text{who}_i] [_s \text{Mary likes } e_i]]]]$

This structure is not well-formed, however, since the C which the *wh*-phrase occupies is not marked $[+WH]$. The *wh*-phrase, therefore, must move again, this time to the first C node in the sentence. Since it is marked with $[+WH]$, the *wh*-phrase stays in this phrase. The resulting structure is that shown in (7).

(7) $[_{\bar{c}}[_{c[+WH]}\text{who}_i] [_s \text{John believes} [_{\bar{c}}[_c e_i] [_s \text{Mary likes } e_i]]]]$

Notice the effect of this rule in structures that appear to have long-distant bindings. That is, where the *wh*-phrase is very distant from its source as in (5), there is actually a local binding by each of the intermediate traces.

THE INTERPRETATION OF ANAPHORS

In programming languages it is important to establish the scope of identifiers in order to interpret them properly. In PASCAL, for example, the scope of an identifier is the block of the program where it appears. This is also an important consideration in natural languages, where sometimes a word acts as a variable (or identifier) and must find its interpretation within a defined scope. In fact, this is another aspect of the logical form of a sentence.

Let us give an example to show how important this notion of scope really is in natural language. Consider the sentences in (a) and (b), where the reflexive *herself* is called an *anaphor*.

(a) *John gave the picture of Mary$_i$ to herself$_i$.
(b) John gave Mary$_i$ the picture of herself$_i$.

Why is it that sentence (a) is ungrammatical whereas (b) is acceptable? In (a) it is impossible for *herself* to be the same person as *Mary*, while in (b) it is necessary. To understand why this is, we need to look more carefully at how the *structure* of the sentence interacts with the *scope* of the anaphor. First, notice how the tree structures for the two sentences differ.

(a)

(b)

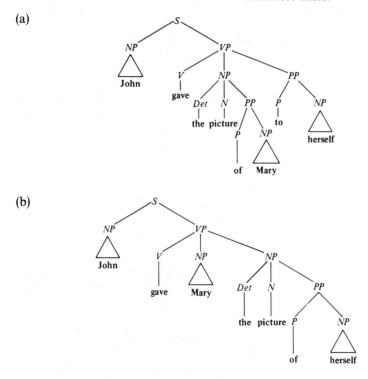

First, let us define the scope of an anaphor as being "the sentence containing it," and within this scope there must be some other *NP* that it receives its interpretation from (the *binder*). But what kind of *NP*? Why can the *NP Mary* be the binder in (b) but not (a)? These sentences are minimally different in one important respect: the relative structural position of the binder to the anaphor. That is, we can define a notion of "closeness" based not on distance in a contiguous string, but on the nearness in a tree. In order to formalize the notion of proximity in a tree, we will make use of a term, *c-command*, defined next.

> *C-Command*: Node α c-commands node β if the first branching
> node dominating α also dominates β.

As an example, consider the following tree. *A* c-commands *E*, but neither *E* nor *D* c-commands *A*.

Now, returning to the examples, we can see that the binder in (b) c-commands the anaphor, *herself*, whereas in (a) it does not. Thus, in the logical form, we now have a way of interpreting anaphors:

> *An anaphor has as its scope the sentence containing it. The anaphor*
> *is bound by a c-commanding NP.*

That is, words seem to affect those phrases that are closest to them in the tree, in the sense of c-command.

In this final chapter we have explored some of the newer developments in the field of linguistics. We have also tried to highlight the many close connections between linguistics and formal language theory as it is studied in computer science. There is a list of readings at the end of the chapter for the reader interested in looking more deeply at the relationship between formal language theory and linguistics.

References for Chapters 8 and 9

[1] Arbib, Michael A., E. Jeffrey Conklin, and Jane C. Hill, *From Schema Theory to Language*. Oxford University Press, New York, 1987.

[2] Chomsky, Noam, "On Certain Formal Properties of Grammar." *Information and Control 2*, 137–167, 1959.

[3] Chomsky, Noam, *Aspects of the Theory of Syntax*. MIT Press, Cambridge, MA, 1965.

[4] Chomsky, Noam, "Remarks on Nominalization," in *Readings in English Transformational Grammar*, ed. Roderick Jacobs and Peter Rosenbaum, Ginn and Company, Waltham, MA, 1970, 184–221.

[5] Chomsky, Noam, *Lectures on Government and Binding*. Foris, Dordrecht, Holland, 1981.

[6] Gazdar, Gerald, "Unbounded Dependencies and Coordinate Structure." *Linguistic Inquiry, 12*, 155–184, 1981.

[7] Gazdar, Gerald, Klein, Ewan, Pullum, Geoffrey, Sag, Ivan, *Generalized Phrase Structure Grammar*. Harvard University Press, Cambridge, 1985.

[8] Hill, Jane and Arbib, Michael A, "Schemas, Computation, and Language Acquisition." *Journal of Human Development 27*, 282–296.

[9] Jackendoff, Ray, *X-Syntax*. MIT Press, Cambridge, MA, 1977.

[10] Kimball, J. "Predicate Definable by Transformational Derivations by Intersection with Regular Languages." *Information and Control 11*, 117–195, 1967.

[11] Myhill, J., *"Finite Automata and the Representation of Events."* WADD Technical Report, 57-624, Wright Patterson AFB, Ohio, 1957.

[12] Rabin, M. O., and D. Scott, "Finite Automata and Their Decision Problems." *IBM J. Res. 3(2)*, 115–125, 1959.

[13] Salomaa, A., "The Generative Capacity of Transformational Grammars of Ginsburg and Partee." *Information and Control 18*, 227–232, 1971.

[14] Schieber, Stuart, "Direct Parsing of ID/LP Grammar." *Linguistics and Philosophy, 7*, 135–154, 1984.

[15] Schieber, Stuart, *Unification Theories of Grammar*. Center for the Study of Language Information Lecture Notes. Stanford University, Stanford, CA, 1986.

[16] van Riesmdijk, Henk, and Edwin Williams, *Introduction to the Theory of Grammar*. MIT Press, Cambridge, MA, 1986.

[17] Woods, William, "Transition Network Grammars for Natural Language Analysis" *CACM 13(10)*, 591–606, 1970.

Symbol Index

Author Index

Subject Index